Theory and Practice of Relational Databases

Theory and Practice of Relational Databases

Second edition

Stefan Stanczyk
School of Computing and Mathematical Sciences
Oxford Brookes University, Oxford, UK

Bob Champion

Richard Leyton

London and New York

First published 1990 by Pitman Publishing
Second impression published 1993 by UCL Press
This edition published 2001 by Taylor & Francis
11 New Fetter Lane, London EC4P 4EE

Simultaneously published in the USA and Canada
by Taylor & Francis Inc.,
29 West 35th Street, New York, NY 10001

Taylor & Francis is an imprint of the Taylor & Francis Group

Printed and bound in Great Britain by
Biddles Ltd, Guildford and King's Lynn

Publisher's Note
This book has been produced from camera-ready copy supplied by the editors.

British Library Cataloguing in Publication Data
A catalogue record for this book is available from the British Library

Library of Congress Cataloging in Publication Data
Stanczyk, Stefan, 1945-
Theory and practice of relational databases/ Stefan Stanczyk. -- 2nd ed.
p.cm
Includes biographical references and index.
ISBN 0-415-24701-2 (cloth : alk. paper) -- ISBN 0-415-24702-0 (pbk. : alk. paper)
1. Relational databases. 2. Database management. I. Title

QA 76.9.D3 S698 2001
005.75'6--dc21 2001027206

ISBN 0-415-24702-0
ISBN 0-415-24701-2

Contents

Preface

First published a decade ago, the ***Theory and Practice of Relational Databases*** gained fairly noticeable popularity, particularly amongst those readers to whom it was primarily addressed - the students. After a decade, however, any book needs reviewing, for the field will have developed, presentation could be improved, choice of topics might be reflected upon and, importantly, the comments from the readers addressed.

Databases evolved into a classic component of computing degrees. The subject became well supported by a wealth of research, exceptional industrial experience and numerous books covering a wide range of topics. However, books on databases run into voluminous proportions and tend to cover the whole spectrum of the subject thus constituting a monographic source of reference rather than being a learning aide.

The book we are presenting now is meant to be just that - a tutorial text that assists the process of learning. It is supposed to have a technological bias, to present the chosen topics in a concise manner, and to incite better understanding through explanations and illustrative examples. In short, the book is meant to retain those features that made the previous edition successful.

Naturally, the book does not aspire to cover all aspects of databases nor does it pretend to present the relational theory in its entirety. The focus is on a coherent, systematic coverage of database design. The primary objective of this book is to present a reasonably comprehensive explanation of the process of the development of database application systems within the framework of the set processing paradigm.

Amongst the rich variety of data models advocated by their authors from time to time, the relational approach has prevailed. However, for applications that require processing of complex data structures the relational approach may not necessarily be advantageous. Application software built around the relational DBMS may require user-defined, complex data structures, appropriate to the domain of that application. Furthermore, certain types of applications do not naturally lend themselves to the relational paradigm. Thus we chose, and not without discussion and controversies, to include a separate chapter covering the object-oriented paradigm as applicable to databases.

Other than that, how different is this edition from the previous one? Well, we have noticed that learning Relational Calculus distracted many students from appreciating the principles of the relational model rather than contributing to its deeper understanding. Relational calculus, being isomorphic to algebra, does not greatly enhance the model *per se*; rather it constitutes an alternative view on its operational part. Accordingly, the chapter on calculus was removed and the exposition of relational algebra strengthened to include the aspects of query optimization and algorithms for algebraic operations.

Furthermore, we have taken the view that relational algebra , being a programming language, deserves a proper software support. Thus LEAP, a straightforward relational DBMS, written by Richard Leyton as an open-source project, has been presented in a separate chapter. The system, equipped with an algebraic interface, can be downloaded from the book's website and used freely by the readers. This brought about an additional benefit - the readers are given an opportunity to study some internal mechanisms of DBMS by inspecting, and possibly experimenting with, the source code - an experience that cannot easily be attempted with any commercially available DBMS.

SQL - by now firmly established as **the** database language - received in this present edition a far more prominent exposition. A reasonably comprehensive description and explanation of the language (largely based on *Programming in SQL*, Pitman 1993) with particular attention to its set-theoretical pedigree is presented; developing more complex SQL programs out of primitive expressions follows. Secondly, an overview of a programming extension is given using as an example PL/SQL - the system developed by ORACLE to support, amongst others, user-defined data types, conditional and looping structures, exception handling, functions, procedures, triggers and packages. Importantly, we have constrained the presentation of SQL to its existing software support; hence no details of constructs specific to say SQL 3 are considered.

The case study gathers all relevant solutions analyzed throughout the book and culminates in a fairly extensive model for a University Information System, partly implemented in **mySQL** and using **Apache** to illustrate the problems and solutions of developing an interface between a database system and the Internet. The choice of software tools was dictated by their free availability, their technical merits and a reasonably high popularity amongst the database developers.

Finally, the book was given a semi-interactive companion, that is its own Internet site **www.theorypractice.org** with references to books, software, discussion groups, research centres and software producers.

In the context of the above amendments, the structure of the book is now as follows:

Chapter 1	Database approach to information systems, the generic 3-level database architecture, characteristics of various database models.
Chapter 2	Entity-Attribute-Relationship as a technology-independent notation for logical data modelling.
Chapter 3	The relational data model introduced as a formal system composed of the structural, behavioural and operational parts. Properties of the model. Relational representation of EAR.
Chapter 4	Relational Algebra and its use for data manipulation. Implementation and optimization of algebraic expressions.
Chapter 5	Description of a didactic DBMS based on relational algebra. Structure, operations and an example of running the system.
Chapter 6	Functional dependencies, Boyce Codd Normal Form. Normalization as mechanism for optimizing the relational structures.
Chapter 7	Multivalued and join dependencies and the relevant normal forms. Selected theoretical aspects of normalization.
Chapter 8	Structured Query Language. Data definition and manipulation. Specifying and programming update transactions.
Chapter 9	Object-orientation as a means for developing databases with inherently complex structures. Object relational database model.
Chapter 10	Procedural extensions to SQL with particular attention to non-relational retrievals and update transactions.
Chapter 11	Case study

We gratefully acknowledge the comments from our colleagues and students, past and present, in the School of Computing and Mathematical Sciences of Oxford Brookes University.

We feel greatly indebted to Dilys Alam and Grant Soanes, our editors at Taylor & Francis. That this book has appeared is a credit to their patience, support and encouragement. We are also very grateful for Alison Nick's sterling work as our Project Manager.

Finally, our thanks to Urszula, Lucille and Frances for them being more than forgiving of our preoccupation with writing this book.

Oxford, June 2001

CHAPTER 1

Introduction

1.1 THE CONCEPT OF A DATABASE

Proper information support is of paramount importance for the management of an enterprise. The successful operation of a road network, a railway system, a bank, a production company or service providers depends on relevant, precise and up-to-date information. The relevant decisions, whether instantaneous (e.g. those taken in real-time production control) or long term (defining strategies or policies, for example), should be made on the basis of multiple facts and these must be properly aggregated, evaluated and analysed in some acceptable time.

Unless the enterprise is small, the task of management is usually divided into a number of coherent functions, such as research and development, planning, production, sales, etc. Each of these functions takes a specific view on the operation of the enterprise as a whole; all of them taken together aim to achieve the ultimate goal - prosperity of the company, successful running of a project, smooth operation of services, or whatever the objectives might be.

Although separately carried out, the management functions are not necessarily disconnected. On the contrary, they affect and influence one another. For instance, financial circumstances determine in some sense planning and production, and limit allocation of resources for research. Production, in turn, determines sales and provides some feedback for research programmes, and so on. Consequently, some decisions made within the scope of one function may overlap with other decisions in some other areas. Also, several managers may use the same data, perhaps differently perceived, aggregated or formatted.

In conventional data processing, each of the management functions is supported by a separate information system. These systems, which operate within some environment (computer hardware, specialized equipment for data collections, specially trained operators), have their own 'private' files and their own 'private' processes developed in a programming language that is most suitable for a particular application.

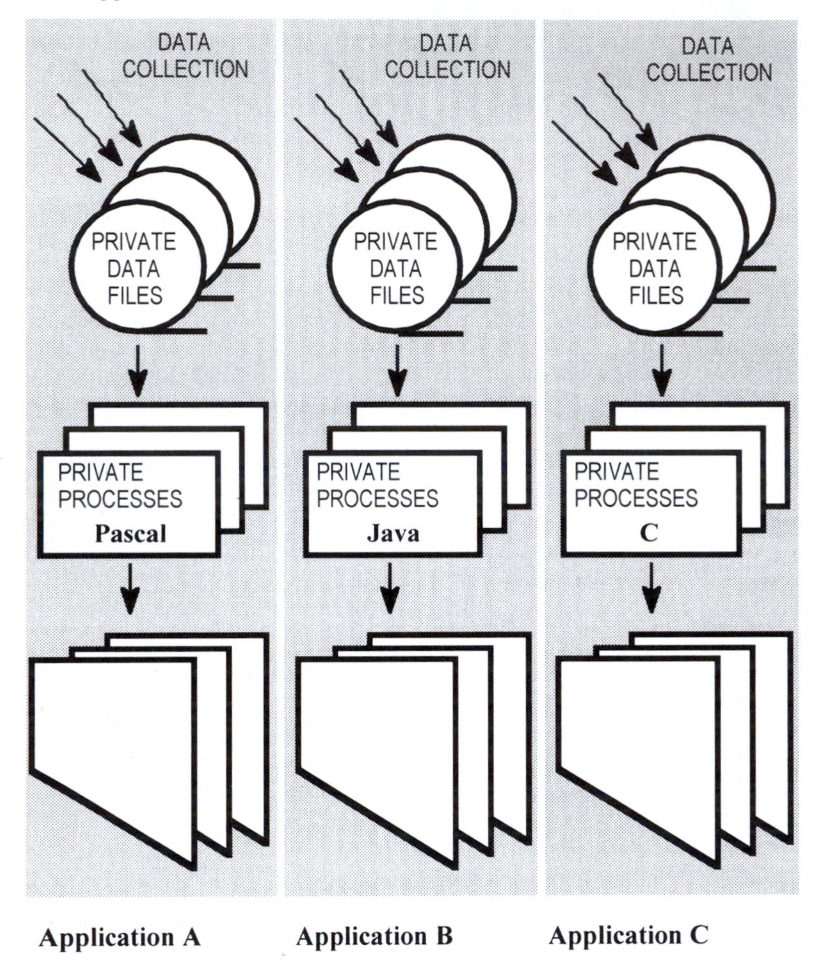

Fig. 1.1 Disjoint information systems

This situation is far from satisfactory. The most commonly appreciated reasons for this dissatisfaction are:

Redundancy of data
Several files contain the same data. The data is likely to be separately collected according to some specific procedure devised for each of the subsystems; a possible use of a sophisticated equipment for data collection makes the whole

process rather expensive. Moreover, the data duplicates are most certainly to be separately updated, thereby involving the risk of inconsistency.

Non-interchangeability of data

Suppose one of the applications is to be extended to incorporate some new functions requested by the users. To produce the required results, this application may need some new data that is not available in its own files but happens to be present in some other system's files. However, due to several reasons - different file organization, different formatting of data, idiosyncrasies of programming languages - the other system's files may not be directly accessible. Hence some additional (and in fact unnecessary) programming effort is needed to convert the relevant files into the form acceptable by the application in question.

Non-interchangeability of processes

Numerous routines are common for all of the applications (sorting, searching, organizing and processing data structures are the prime examples), yet they must be coded separately, according to the specific programming languages' requirements. Again, some waste of programming effort occurs.

Non-transparency of the application software

A considerable part of the application software handles purely data processing matters and this conceals the application logic rather than bringing it out. It is, then, rather difficult to reconstruct the application logic by reading the relevant code - these two types of information are expressed at completely different levels of abstraction.

Inflexibility of the application software

The application software (which essentially represents processes, not the data) contains some built-in knowledge about the data (such as data types and range of variables, for example). This knowledge is duplicated in every program that uses the relevant data and makes the global data consistency control difficult. Moreover, should these types, ranges etc. change (for whatever reason - extending a field size and incorporating a new field into a record may serve as a typical example) considerable reprogramming must necessarily be done throughout the whole application software.

Uncontrolled expansion

There is no mechanism to control in any systematic way a possible (and likely) growth of both the data and the processes, neither is there any form to balance the conflicting requirements. Inevitably new data, new collection and updating procedures, and new processes will be added to the systems, thus making the system programming support and resource allocation increasingly difficult.

To summarize, the management of the enterprise is not supported by any coherent method for corporate control of the data. Yet the data is one of the enterprise's assets, just as valuable as human resources, buildings, machines and finances are. *The database approach to information systems provides the management of the enterprise with means to impose centralized control over its operational data.* This is the main advantage (and indeed, the objective) of having a database system implemented.

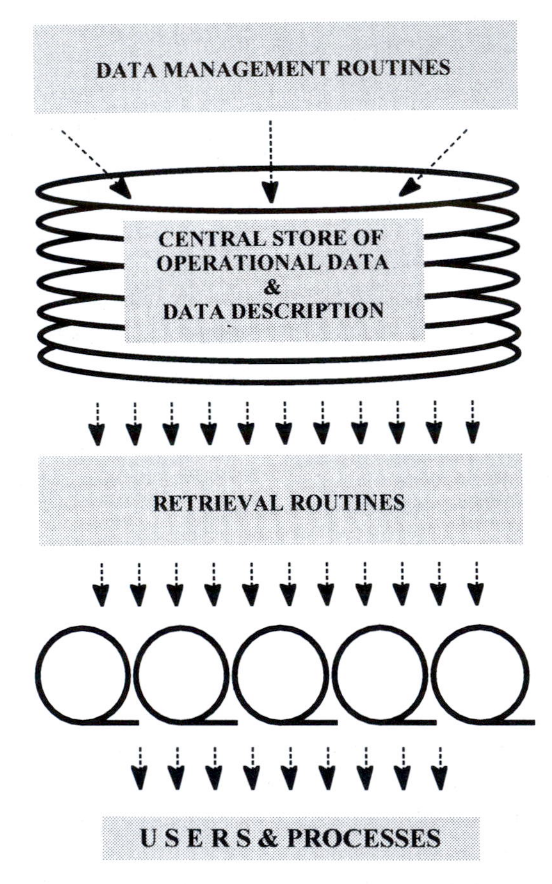

Fig. 1.2 A simplified database structure

A general concept of a database is depicted in Fig. 1.2. We shall give a detailed account of its structure shortly, but at this point we can view a database as a structured collection of operational data together with a description of that data.

The heart of the database system is then a central store of data - an integrated collection of records with any excessive redundancy eliminated (some duplication may occur for e.g. validation purposes). The data is shared among all the users of the system be they casual interrogators, application programmers (or programs themselves) or the *Database Administrator* (DBA).

The DBA (a team rather than a single person) can be thought of as a supreme controller who supervises every aspect of the database existence. In particular, the DBA is responsible for the database information content, the security and integrity of data, the storage structure and access strategy and for monitoring the performance of the database - making necessary adjustments whenever necessary.

All communication between the physical representation of the data and any user is done through the *Database Management System* (DBMS). This means that virtually every activity in the system (including defining and modifying database structures, inserting, deleting and updating values, and all kinds of retrievals) is controlled by the DBMS.

The DBMS contains a variety of facilities including a *data definition language* (to create and modify the database structures - files, users and their privileges), a *query language* (which supports all forms of retrieval and updating) and numerous interfaces to liaise with the operating system, telecommunication system, programming languages and other utility software. It also contains data validation routines and maintains a *Data Dictionary* - a complete description of the database structure and content.

1.2 DATABASE ARCHITECTURE

The database architecture whose brief account is the subject of this section was proposed by the ANSI/X3/SPARC group (Tsichritzis, 1978) in an attempt to provide a general framework for database systems, quite irrespective of their underlying data models (hierarchical, network or relational).

The database architecture (see Fig. 1.3) essentially comprises 3 levels - *conceptual*, *external* and *internal* - in an attempt to separate the logical and the physical aspects of the system. The main idea is to provide a framework that makes it possible to consider the data separately from processing and to insulate the data from all implementational aspects, be they hardware constraints, or software facilities, or whatever.

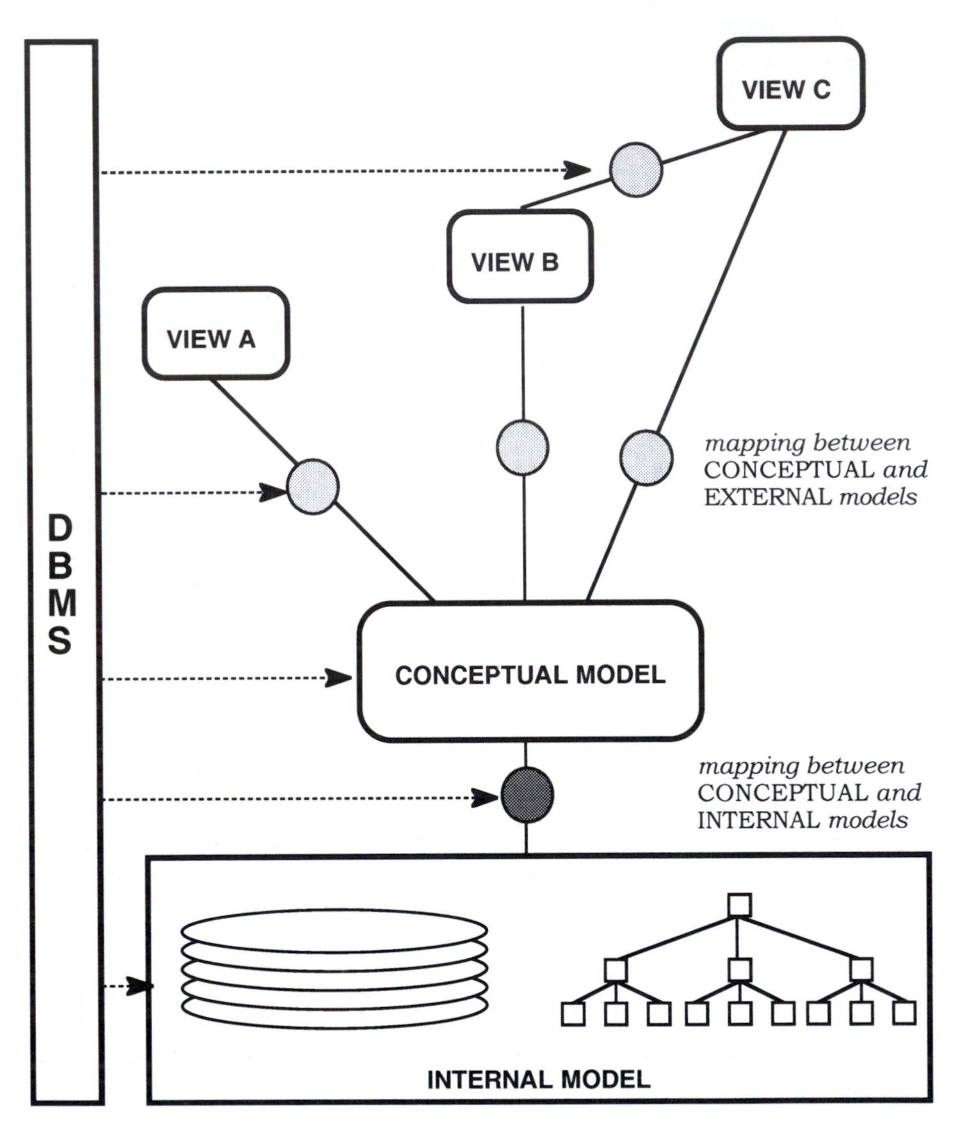

Fig. 1.3 The ANSI/X3/SPARC architecture for a database system

The **conceptual model** is a common, unconstrained view of the data. It is a model that contains all the relevant (to the information system being developed) facts recorded in some suitable notation. At this point it is immaterial how this data is going to be processed or stored; all that counts is its relevance and truthfulness. The conceptual model is supposed to be a true image of the Real World as perceived by all parties concerned - the users and the developers alike.

Since all the data in the database is integrated, only a relatively small portion of it is of interest to a particular user . We call this portion of the data a **view**. There can be many separate or overlapping views according to the specific user's requirements. The views can be created or destroyed as circumstances dictate, hence the whole structure of views is dynamic.

The **internal model** represents the actual storage representation of all the data in the database.There is obviously just one internal model and it is closely connected to the actual software facilities provided by the computer system on which the database is implemented.

All the models are recorded (stored and kept up-to-date by the DBMS in both the source and the object form) in terms of a Data Sub-Language (DSL) as **schemas**. The *conceptual schema* comprises definitions of all the logical units of data together with their types, the logical relationships among them and the appropriate validation procedures. The conceptual schema does not address the questions of storage structure and access strategy in any way; although written in DSL it does not depend on any particular programming language.

Every view is described by means of an *external schema* (also stored by the DBMS). It contains descriptions of each of the various types of external records which are defined on conceptual records but not necessarily in a one-to-one correspondence. The *internal schema* (again stored by the DBMS) defines the structure of the internal records and contains information on possible indices, applicability of field values for hashing or indexing and similar properties or physical relationships.

The mappings CONCEPTUAL ↔ EXTERNAL and CONCEPTUAL ↔ INTERNAL (both of them stored by the DBMS, of course) ensure the database model coherence and facilitate data independence.

The notion of **data independence** is fundamental to the database theory. It gives the DBA the freedom of changing both the physical and the logical aspects of the database system without disturbing the applications built on the database.

The CONCEPTUAL ↔ INTERNAL mapping supports the **physical data independence** - *a measure of the immunity of an application to changes in the storage structure and access strategy* [Date 00]. The question of how the data is actually stored and accessed is immaterial from the application viewpoint, yet in conventional data processing the knowledge of these physical aspects of the data organization is built into the application logic, and consequently constitutes a major part of the application code. Should then any change occur (e.g. a direct hashing organization is to be replaced by a B*-tree indexed one, say for query performance reasons), the majority of application programs will require a substantial re-development, despite the fact that the change is purely circumstantial and has nothing to do with the application logic.

A similar change in a database system would indeed necessitate some redefinition of the CONCEPTUAL ↔ INTERNAL mapping, and perhaps some modifications in the DATA DICTIONARY but the applications would remain undisturbed.

The CONCEPTUAL ↔ EXTERNAL mapping supports the **logical data independence** - *a measure of how well an application view is insulated from changes in the conceptual model of the database* [Date 00]. Any increase of the data will result, of course, in some changes (and perhaps restructuring) in the conceptual model - and the corresponding changes in the mapping. The existing applications, however, will not be affected, unless a planned development of a particular information system makes use of a wider range of the data available in the modified database.

In summary, the 3-level database architecture provides a universal framework for system development clearly separating the logical aspects of database design from the technological means of implementation. This architecture is universally accepted as a reference base (at the highest level of abstraction) for all databases irrespective of their underlying logical data models, programming systems and other technological means.

1.3 LOGICAL DATABASE MODELS

This book deals almost exclusively with the relational model - so much so that even in the chapter that describes the foundations of object databases references to the relational theory are continually being made. There is a good reason for that approach. The relational model, rigorously derived in 1969-70 (Codd, 1970) as a formal system equipped with transparent and compact notation, transformed into a high-level practical programming system, and successfully used for application system development has dominated research, development and industry for some 30 years now.

However, the relational model is not the only one that application systems have been built upon. Prior to the relational dominance two database models gained some (relatively short-term) acceptance - the models that followed the hierarchical and the network (CODASYL) approaches. Neither the former nor the latter could rightfully be considered as a model per se (not in the sense the relational model is, anyway) and neither of them enjoyed any specific mathematical support. In fact, each was an extension of a particular programming system with a set of guidelines somehow abstracted from the accumulated experience and guarded by the relevant international standards.

The underlying data type for the hierarchical model was that of hierarchies of trees but many applications developed as hierarchical systems tended to force further extensions in the form of multi-threading or imposition of additional ordering. The CODASYL model, in turn, was defined in terms of *ordered* sets that were describing a relationship between the master record (the 'owner' of the set) and the subordinate records (the 'members' of the set) . The relationships between sets were contrived via link-sets (intersections) thus the structure could be seen as a network (i.e. a graph); hence the name.

Although many DBMSs developed to support either approach comprised high-level programming constructs to manipulate the relevant data structures (at the slightly higher level of abstraction than procedural languages used at that time), much of the data semantics was, as of necessity, built into application software. Furthermore, data definition and manipulation involved the use of pointers and application software required explicit navigation. In some systems the physical data independence was not necessarily adhered to; the logical data independence was not supported at all. Consequently, only very precisely specified applications with stable conceptual data models (and hence no need for restructuring) could be (and were) built.

Shortly after Codd's announcement of the relational model numerous attempts to use other mathematical concepts for database field were investigated. Since relations may also be perceived as predicates of mathematical logic, a deductive database model was intensively researched; this gave rise to the development of knowledge-based systems where the underlying database (or rather knowledge base) comprised not only the 'raw' data but also held some inference rules and a language capable of processing both. Correspondingly, applications could be equipped with classical database functionality **as well as** facilities for conjecturing facts, reasoning about the results of processing and the way that processing was done.

Another interesting approach relied on the concept of (multivalued) functions. The foundations of the functional model were laid down by Abrial (1974); the paper defined foundations for a universal model which was capable of storing and interpreting facts, rules, assertions, inferences, etc by utilizing two primitive notions: *categories* interconnected via *access functions* (relationships between categories), which themselves could be characterized by compositions of categories. From these two primitives binary relations were built that represented a model of the real world.

The binary relational model eventually evolved into the extended functional model where *base functions* (whether single or multi-valued) represented stored data while *derived functions* (various combinations of inversion, composition, recursion, transitivity, etc.) represented processes. Possessing a very elegant notation and a transparent programming language (functional, naturally) the model never materialized in any real industrial product, however the model's cognitive values are unquestionable.

It is fair to say that the overwhelming majority of application systems are now built as relational databases. Whatever the virtues (or otherwise) of the pre-relational systems were, neither hierarchical nor network DBMSs are in active use anymore. Whatever the strengths (or otherwise) of post-relational systems will be, time will show.

CHAPTER 2

Data modelling

2.1 MODELLING THE REAL WORLD

Modelling is an activity whose aim is to produce a correct, complete and consistent representation of the Real World - or more precisely that part of the Real World which is of interest to the designer of the target Information System. This representation must meet a number of criteria; apart from being a true image of the Real World it must be comprehensible by both the user and the designer, and must be implementable in a certain environment.

The complexity of the Real World can hardly be expressed even in a natural language, such as English, French or German. But observations, reasoning, assumptions and other forms of intellectual speculations have to be expressed in this way; after all, we do formulate and record our experiences somehow. However, the natural language fails the criteria. While rich and flexible, it allows ambiguity and inaccuracy, it is not transparent enough, it cannot be subjected to the uncompromised rules of computational logic - in short, a representation of the relevant part of the Real World expressed in natural language may not be implementable in the computing environment.

Strictly speaking, developers create typically two models - an Informal Conceptual Model (ICM) and a Computerizable Conceptual Model (CCM). The ICM is a developer's perception of the Real World, a kind of a mental image created on the basis of observations, analysis, interviews, personal knowledge, and similar factors. Very often, this image never gets written or otherwise systematically recorded in its entirety. On the other hand, the CCM is a result of the analyst applying certain recognized methods and techniques, and typically is a part of the documentation of an Information System.

While transformation of the ICM - ultimately seeming coherent, consistent and complete - into the CCM can be done in a fairly rigorous manner, the transformation from the informal thoughts, words and actions of the Real World to a systematic form of ICM is not. This transformation, itself unspecifiable and inexact as done by humans, is disturbed even further by the influence of some questions deeply rooted in people's minds: the questions from which they are quite unable to separate themselves while doing this transformation despite, perhaps, their awareness that these would have to be considered either later or at least separately.

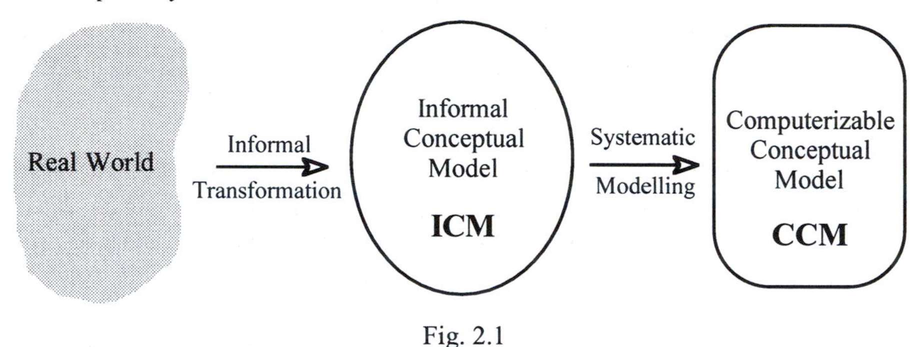

Fig. 2.1

Among others, there are matters related to the choice of strategy and reasons for which the system is being created, and to the choice of subset (i.e. which phenomena, objects and actions are to be included in the model). These are, of course, valid questions. But neither the relevant solutions nor the actual method to find them should affect the way the Real World is modelled.

A method for modelling the Real World must provide a device to express any model in a way that is comprehensible by people and machine-processable in the environment in which this model is to be implemented. At the same time the model must be isomorphic to the relevant part of the Real World; that is the constructs and actions occurring in the model must be in one-to-one correspondence with the objects and their behaviour in the Real World.

We need, then, a notation to transform vague descriptions and informal thoughts and words, i.e. Real World actions, to a systematic form, perhaps with a restricted representation, but such that it is accurate, complete and consistent. Otherwise, the implementation of the target Information System would behave in an unpredictable manner!

Typically, only certain aspects of the Real World get represented in its corresponding computational model. The aspects that are (or seem) unnecessary for the target system objectives are disregarded. For example, recording certain characteristics of people that refer to their physical appearance (such as height, weight, etc.) seems irrelevant (if not unlawful) in a system that supports task allocation in a company while the same attributes are, of course, of importance to a health care database.

Furthermore, the representation of such a restricted Real World is typically discrete and most likely finite, for the computational processes are discrete and finite. Thus the continuous Real World is replaced by a number of distinguishable objects together with their interactions, associations or relationships and the processes that transform these objects.

The fundamental activity in modelling is then the ability to distinguish objects that are relevant to the functions of the target system, the functions themselves being typically modelled as processes. How can we distinguish objects both from the environment where they actually occur and from each other ?

To be able to distinguish anything one has to specify the features that make a *thing* different from *anything else*; thus one has to characterize an object by specifying its properties. Often we may have independent knowledge about the existence of a particular object, but equally frequently a set of properties will **define** an object for the sake of the implementation, for that object may not physically exist in the Real World - yet its existence in the object system may be desirable.

2.2 ENTITY - ATTRIBUTE - RELATIONSHIP MODELLING

The essence of the database approach to information system design is that the underlying database must support in principle any application - whether devised at the very moment of the database design or at some later stage. Therefore the data structures that hold the data in the database must by easily modifiable. That is, structure modification (for instance adding a new kind of data) must not affect the already implemented operations. This question of data independence from processing will be addressed in Chapter 3.

For some time now the Entity-Attribute-Relationship approach (EAR) to data modelling has gained some popularity, specifically in the context of relational databases. The method was introduced by P. P.Chen in 1976 and since then has been much developed and even computerized. Here, the method is presented in a rather simple way, primarily because it is regarded as a tool in designing relational databases. Readers interested in the method in its whole complexity are directed to the original paper (Chen, 1976); very detailed accounts can also be found in Tsichritzis (1982).The method stems from perception of the Real World through finite objects. The objects are essentially of three kinds: entities, their attributes and the relationships among entities. Thus attributes describe entities, which in turn are associated by relationships.

Definition 2.1

Anything that has reality and distinctness of being in fact or in thought can be considered as an ***entity*** *; alternatively, an entity is a physical or abstract object that exists and can be distinguished from other objects.*

Example 2.1

John Wilkes with a student number 40079663 is an entity since the distinct features described by (**name, student-number**) uniquely identify a particular person existing in the universe.

Example 2.2

A Ford Escort car with a registration number KGX 601Y is an entity since it uniquely identifies one particular car.

Example 2.3

By contrast, a book on databases published in 1989 is not a correct representation of an entity - the features do not identify any particular book (but rather a set of books on databases published in that year).

An entity can physically exist (such as car, person, book, part, building, etc.) or may be an abstract or a concept.

Example 2.4

BA Flight# 897 to New York is an entity since its distinct features (**flight#, destination**) distinguish one particular flight from many other flights (whether to New York or somewhere else) that occur in the universe.

Example 2.5

A course 8049 on Database Design taught by Dr Brown in the summer term is an entity since (**8049, Database Design, Dr Brown, Term3**) identifies uniquely a particular course from all the courses taught in a university.

The entities (or, more precisely, entity occurrences) are then characterized by features, i.e. some particular values of the properties described by ***course name***, ***student number*** or ***flight destination*** in the above examples.

Definition 2.2

> ***Property** is a named characteristic of an entity.*

Properties then identify entities and - as we shall see later - allow one to classify entities and to relate them in some manner. An entity may possibly be described by a great number of properties, not all of them being necessarily of importance to the target system.

Definition 2.3

> *Properties that are to represent an entity in the target system (i.e. those whose values are to be stored in the database) are called **attributes.***

Definition 2.4

> *For each attribute there is a set of all permitted values called the domain of that attribute.*

It is important to distinguish between an ***attribute name*** and an ***attribute value*** (in a way this distinction is similar to that between *variable name* and a *variable value* in programming languages). For example, an attribute *flight-number* (name) may draw its values from a domain whose elements (i.e. permitted *values* for this attribute) are: {BA701, AF345, LO242, TWA775}. Entities (or, more precisely again, entity occurrences) that are of the same type can be considered as belonging to the same class. For instance, (**Ford Escort, KGX 601X**) and (**Austin Maestro, A556 VVV**) are particular occurrences of the class *passenger-car*.

Definition 2.5

> *An entity type is a class of entity occurrences characterized by the same attributes.*

In this context, an entity type can be regarded as an abstraction of a class of entity occurrences, that is a higher-level object that represents all its possible occurrences.

A method is appreciated better if it contains a suitable graphical notation capable of recording design decisions and of showing the results of its actions. Consequently, the results of an EAR analysis are often presented in a graphical form - ultimately more transparent and more easily comprehended by both the user and the developer. A standard form of graphical representation is shown in Fig. 2.2.a. Though many conventions exist, usually an entity is represented by a rectangularly shaped box, while attributes are given in angled boxes connected by straight lines with the featured entity. Two alternative representations (attribute names listed by the side of the entity box and linear record of entity name followed by attribute names), both particularly convenient for large data models, are shown in Fig. 2.2.c.

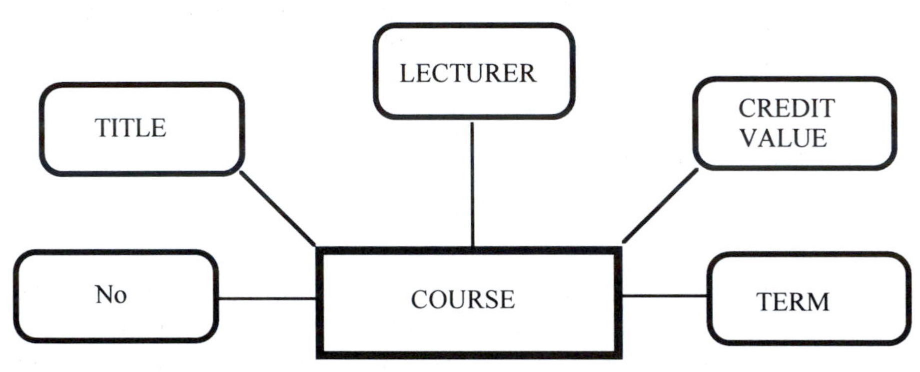

Fig. 2.2.a Graphical representation of the entity type COURSE

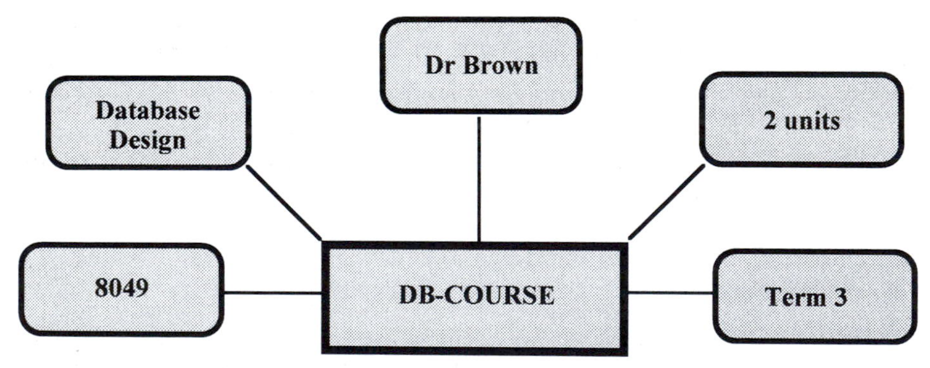

Fig. 2.2.b Graphical representation of an occurrence of the entity type COURSE

Fig. 2.2.b has been given only to underline the difference between a type and an occurrence. The EAR modelling is essentially a type-driven method; the Real World is perceived by means of abstractions or types rather than particular occurrences. In this sense, the EAR conforms to the principle of abstraction that allows developers to build up solutions without taking into account intricate details, subtleties or particularities - these are typically dealt with at the subsequent stages of technical design.

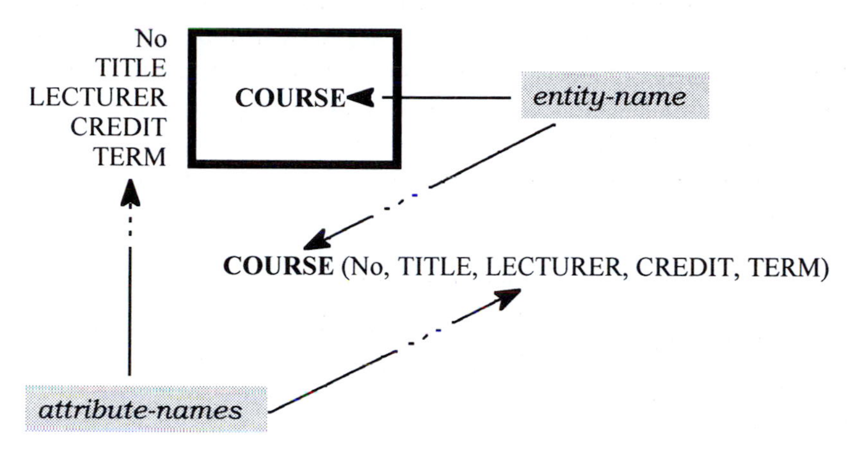

Fig. 2.2.c

We have already noticed that numerous properties could characterize an entity - the developers then select those relevant to the target system objectives and functionality. Entity integrity plays an important role in guiding the selection of the relevant attributes. The principle says that every entity occurrence of an entity type must be identifiable by the values of its attributes and consequently the number of attributes must be large enough. For instance, (**Rover Mini, Chassis#990876533A12**) and (**Rover Mini, Chassis#990656442C02**) clearly represent two different occurrences of the entity type *passenger-car*; had we taken only a single attribute *make* (e.g. **Rover**) to describe this entity, the above two occurrences would have been indistinguishable! (and therefore would constitute one and the same occurrence).

Not necessarily *any* set of attributes form an entity that represent an object from the Real World either. Consider for example an entity that is supposed to represent bridges. Naturally, some properties, (such as length, width, span#, capacity) are common for all types of bridges. However, there are attributes applicable only to certain types while other types do not have those properties (e.g. cable characteristics apply only to the suspended constructions but not those made of steel). An entity that would comprise of all these attributes would be a result of *over-abstraction*, that is it would constitute an attempt to represent all different types by a single entity. Hence some attribute-values would be null and other properties which their real world equivalents do possess may not be recordable at all. In other words, the entity would contain separate groups of occurrences, each group representing a different object type.

There are cases where a hierarchical structure of entities is legitimate, for the Real World objects form that hierarchy in a natural way, and capturing this in the data model might be beneficial. Consider, for example the notion of University membership (see Fig. 2.2.d). Whether junior (STUDENT) or senior (LECTURER), every member (super-entity) is described by attributes common to both types (identification number, name, birth date, permanent address) while differentiating attributes are applicable to one type but not the other. This type of modelling thus assumes, of course, that no member can be classified as both the student and the lecturer.

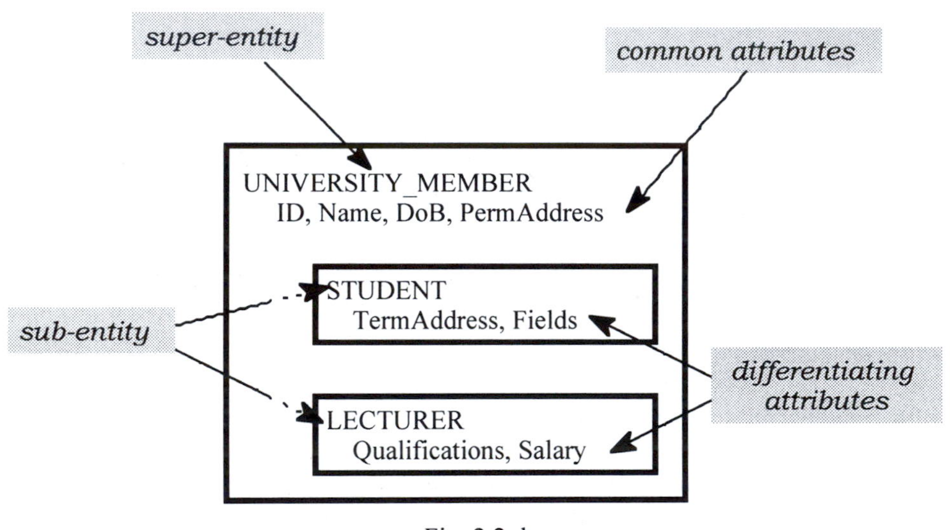

Fig. 2.2.d

Modelling through super/sub-entities presents some advantages particularly in situations where different types of relationships can be identified between those and the other entities within the data model. Ultimately, the concept of mutually exclusive sub-entities forms one of the foundations of Object-Oriented (or, more precisely Object-Relational) databases which will be considered in some detail later in this book.

Having assumed that any two occurrences of an entity type can always be distinguished by specifying all their attribute values, the question now arises - do we need *all* the attributes to make this distinction? From a database perspective such referencing would be rather inconvenient, cumbersome and time consuming. What we are looking for is a *minimal* set of attributes that would identify a particular entity occurrence equally well.

Definition 2.6

> *An attribute (or group of attributes) that uniquely identifies every entity occurrence within its class is called a* ***candidate key.***

Note that in case of a compound key (that is one composed of a number of attributes) the requirement of minimality (smallest possible number of attributes) is in force.

It is possible that an entity type does have more than one candidate key. In such cases one of them is designated by the database designer as the **primary** key, i.e. as the primary means of identifying entity occurrences; all the others are **alternate** keys.

Example 2.6

Identify possible candidate keys in the entity defined as follows:
STUDENT (NAME, Id-No, ADDRESS, BIRTH-DATE, SEX, STUDY-FIELD)

Possible candidate keys include:

Id-No typically this is a unique number assigned to every student and never re-assigned to anybody else

(**NAME, BIRTH-DATE**) note that NAME alone would not be a candidate key since two (or more) different students may actually have the same name; on the other hand one could argue that two persons, each with a name John Smith, could have been born at the same time; adding an additional attribute, say ADDRESS, would probably resolve this.

It is worth mentioning that (**Id-No, NAME**) is a superkey rather than a key since it contains a candidate key as its subset.

Note that entity types need not be disjoint. Indeed, a particular entity occurrence can belong to two or more classes. For example, (**John Smith, 22-May-69, High Street**) may denote an occurrence from a class *student*, but it may well be an occurrence of the entity type *bank-customer*. In some circumstances both entity types can exist alongside each other within an information system though they might need different keys.

Having distinguished the objects and modelled them in the form of entities we need to describe the way they interact or are associated with each other. This is a kind of natural consequence of our breaking the continuum of the Real World into separate objects. A generic word - ***relationship*** - is commonly accepted as standing for all semantically possible kinds of associations or interactions between entities; details and nature of a particular relationship are then described by its name and by a further description, if necessary.

A relationship involves essentially two or more entity types. Intuitively speaking, an entity occurrence from one class may be associated with an entity occurrence from another class though quite often there is a need to define associations among different occurrences of the same entity type (these are so-called co*nvoluted relationships*). If this property is generally valid throughout both classes, then a relationship is said to exist between the corresponding entity types.

Definition 2.7

*A **relationship** is a named directed mapping between two entity types. The three kinds of mapping are defined as follows:*

- ***one-to-one***
an element from one class is mapped to exactly one element in another class

- ***one-to-many***
an element from one class is mapped to one or more elements in another class

- ***many-to-many***
zero, one or more elements from one class are mapped to zero, one or more elements in another, and conversely, the elements from the second entity class are mapped in the first one in an exactly identical way.

Whatever the kind, if a relationship is defined on *all* elements of a particular class, the relationship is said to be **compulsory** on that class, otherwise it is **optional**.

We will consider different kinds of relationships using as an example a university information system for monitoring students' progress and showing the corresponding graphical notation.

Case study

The following scenario will constitute a base for the data modelling (and, later on, database design) for a University Information System. Students at that University read for a degree within the framework of a modular system whose principles are as follows. The University through its departments provides a number of modules, each module being characterized by its code, title, credit value, module leader, other teaching staff and the department they come from. A module is coordinated by a module leader who shares teaching duties with one or more lecturers. A lecturer may teach (and be a module leader for) more than one module. Students are free to choose any module they wish but the fundamental three rules must be observed:

- some modules require pre-requisite modules,
- every student carries out a project supervised by one lecturer,
- the degree titles are defined by the choice of modules.

The database will also contain some personal details of students and staff.

The following entities can initially be identified:

- physical: **STUDENT**, **LECTURER**, **OFFICE**
- abstract: **DEPARTMENT**, **MODULE**, **PROJECT**, **RESULTS**

the last one being of an association type, i.e. a *weak* entity that would not exist without reference to some other entities within the model.

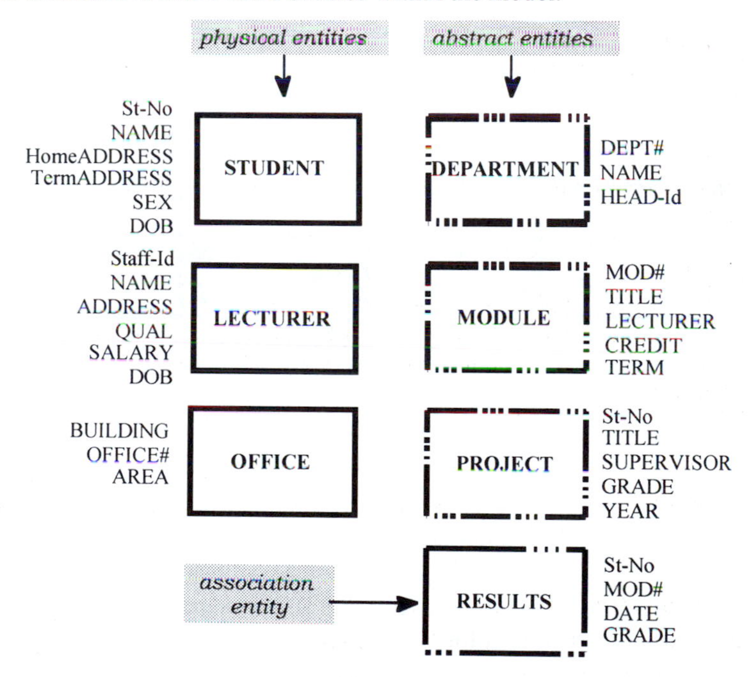

Fig. 2.3

We will now analyse in detail the relationship between some of the above entities. From the specification of the problem we can infer that the relationship between **STUDENT** and **PROJECT** is of a ***one-to-one*** type. Furthermore, any student must undertake an individual project, i.e. no two students can do the same project and no student does more than one though some possible projects have not been chosen by any student. Fig. 2.4 renders precisely the above meaning.

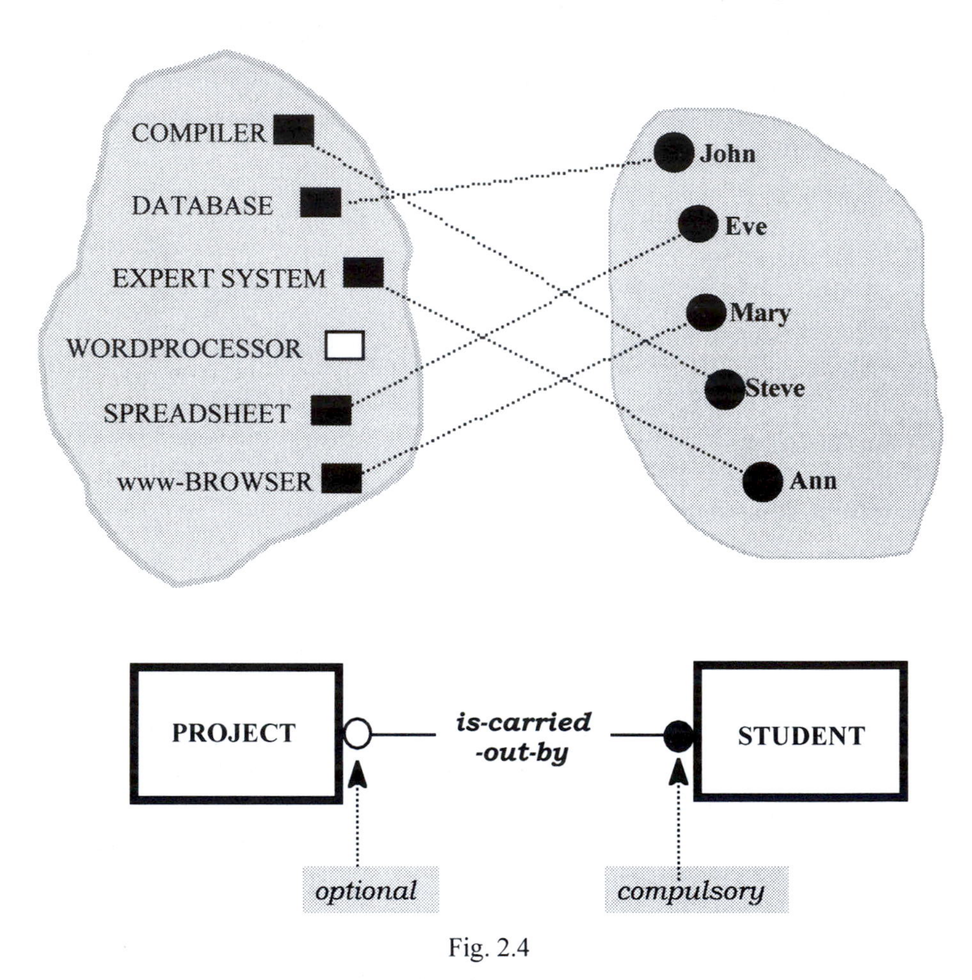

Fig. 2.4

Fig. 2.5 represents a ***one-to-many*** relationship between **LECTURER** and **STUDENT** (in the sense of personal tutorship). A member of staff **may** be a Personal Tutor for a number (none, one or more) of students whereas all students do have a personal tutor. It thus follows that not necessarily all staff are engaged in counselling.

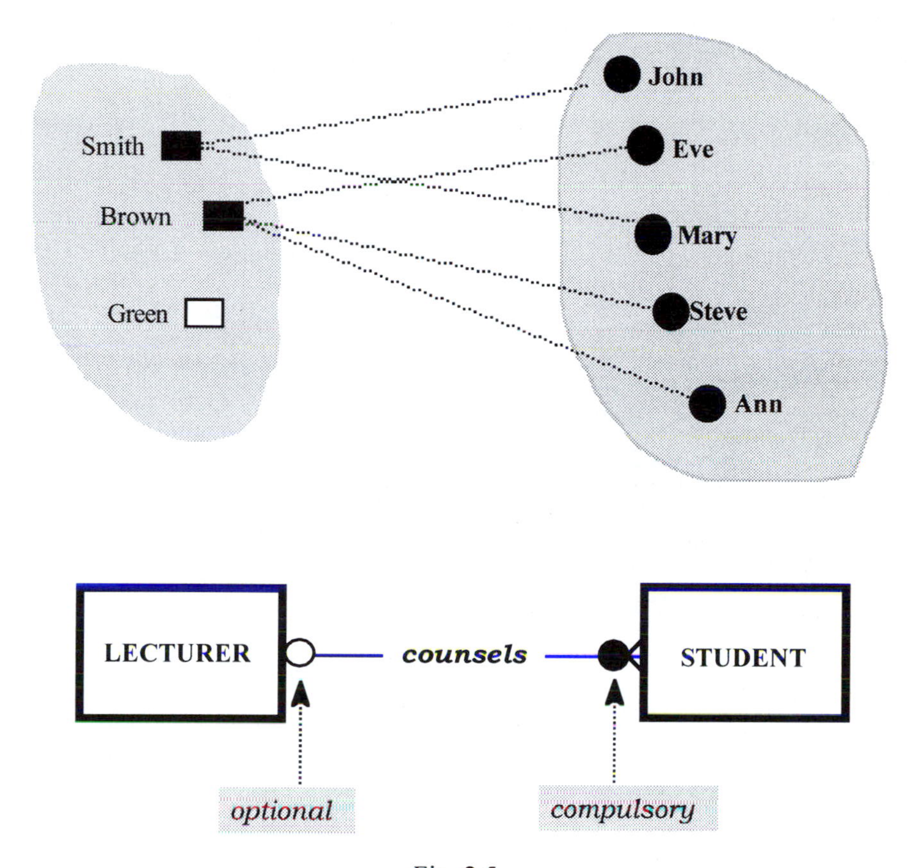

Fig. 2.5

Similarly, Fig. 2.6 represents a ***many-to-many*** relationship between **STUDENT** and **MODULE**. As it happens, some modules attract many students and some students sign up for many (that is, at least one possibly more) modules; some modules may have been unattended, though (say a newly developed module in the middle of a term).

However, a *many-to-many* relationship does present a problem in terms of representing it by structures available within a particular database model (whether relational or otherwise). The corresponding structures would contain a substantial amount of redundancy (since many occurrences of one entity would have to be associated with some occurrences of the other) and, consequently, they would exhibit undesirable update properties. Hence, the relationship is typically decomposed and represented by an association entity linked via two *many-to-one* relationships to the original ones (see Fig. 2.7).

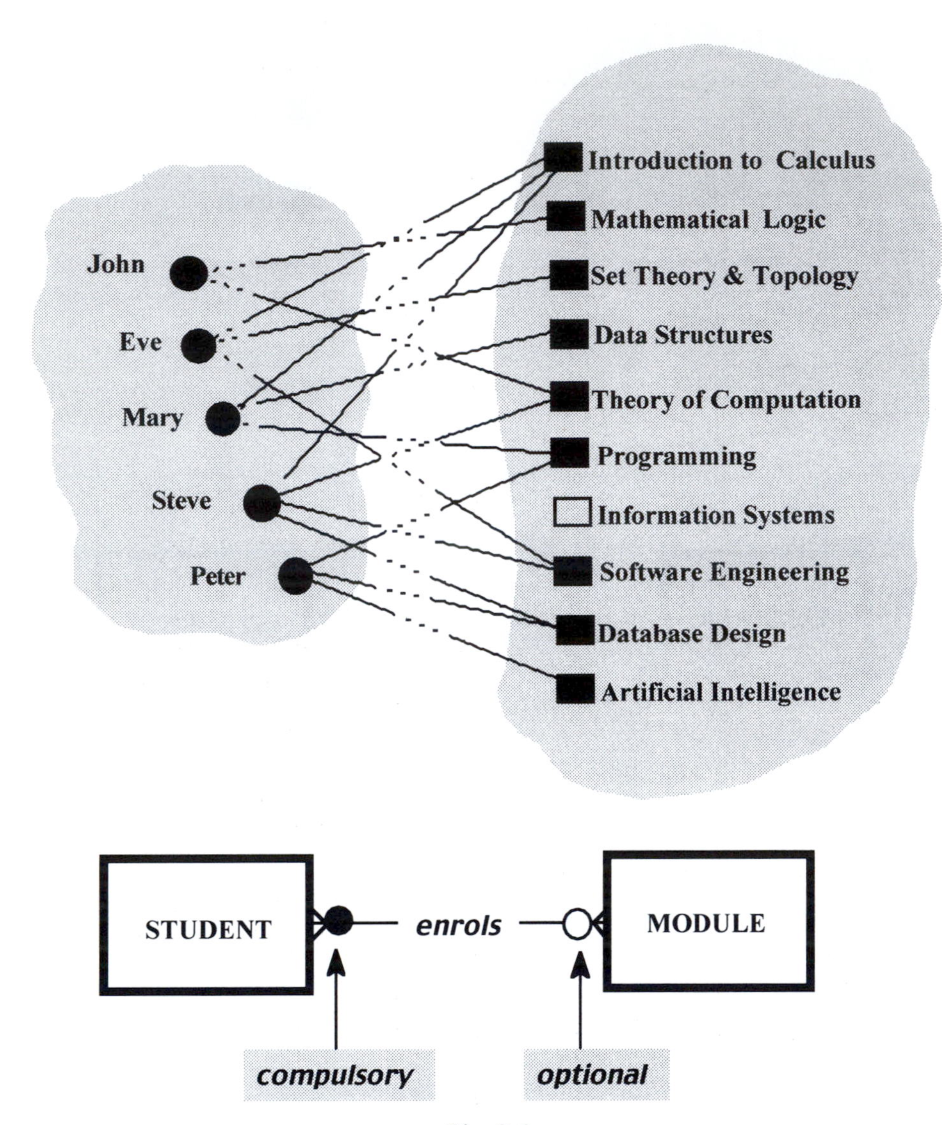

Fig. 2.6

Having considered all possible ways in which the entities are associated we could now produce the global EAR schema showing the entire image of the information content (in a somewhat abstract form) to be stored in a database. A possible form of such a schema is shown in Fig. 2.8.

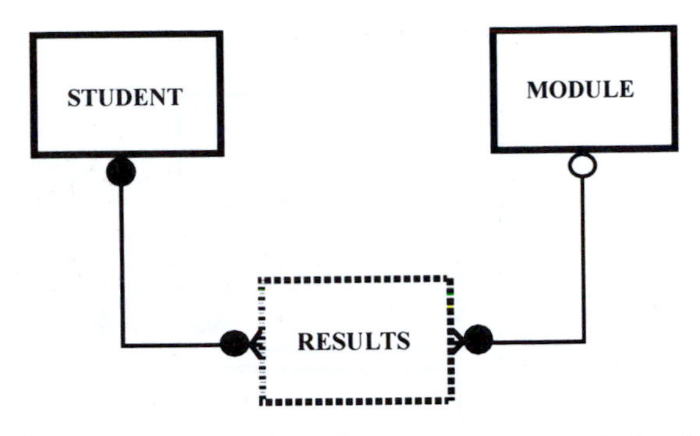

Fig. 2.7 Decomposition of the *many-to-many* relationship

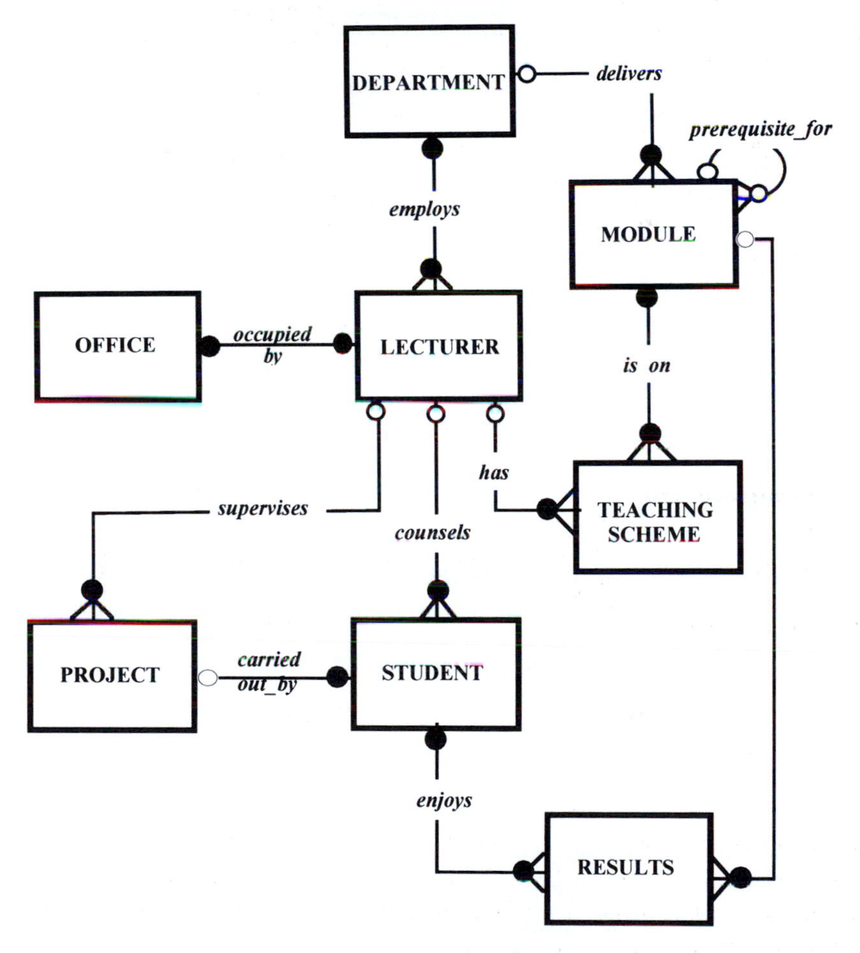

Fig. 2.8 Global EAR schema

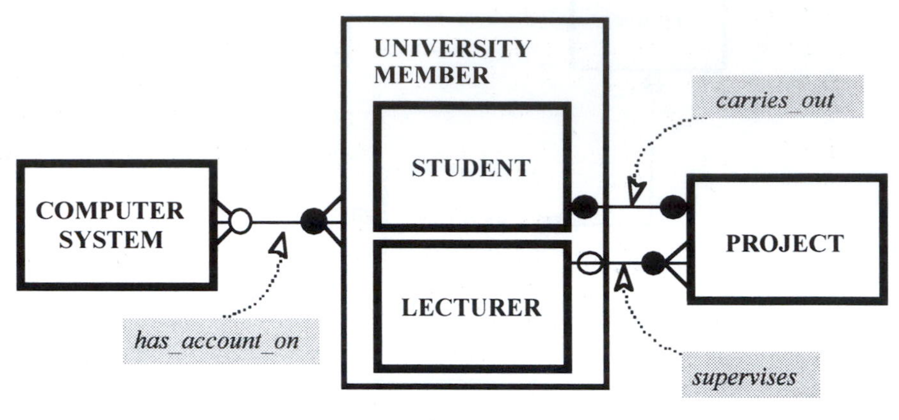

Fig. 2.9

At this point we may consider potential benefits from modelling through the concept of super/sub-entities. For cases where a particular super-entity (say UNIVERSITY-MEMBER) is associated with another entity within the model via one type relationship but sub-entities are in various types of relationships with some other entities the data model becomes richer for some additional meaning (essentially part of processes occurring within the system) is added thus making the perception of the Real World more accurate. An example of such a model is presented in Fig. 2.9.

2.3 EXERCISES

For the problems described below identify entities and their attributes, describe relationships among entities making the necessary assumptions (whenever needed) and draw the global EAR schema.

2.1 Consider the EAR schema presented in Fig. 2.8 and 2.9. Describe precisely the the meaning of relationships and state the assumptions which could not have been inferred from the initial specification on page 21.

2.2 Corporation for Rescuing Automobile Stake Holders (CRASH) provides repair repair services for its members whenever they are in trouble with their cars. While the costs of all repairs are registered, the services are not charged to the members except for surcharges on replacement of some parts (e.g.

windscreens). Anybody may become a life-member of CRASH upon recommendation of another member of the corporation and payment of a rather high fee. A member may request a service at any time and for any car (whether owned or not). The cars are of such a good quality that they do not break down more than once per day. The engineers employed by CRASH are excellent and have all the necessary tools, parts and skills to repair any car at any time.

The operations of the corporation are supported by an information system whose underlying database stores the details of members, engineers, cars that have ever been dealt with by CRASH and all repairs undertaken by the Corporation.

2.3 A relational database is to be designed for a company that deals with industrial applications of computers. The company delivers various products to its customers ranging from a single application program through to complete installation of hardware with customized software.

The company employs various experts, consultants and supporting staff. All personnel are employed on a long-term basis, i.e. there are no short-term or temporary staff. Although the company is somehow structured for administrative purposes (that is, it is divided into departments headed by department managers) all projects are carried out in an inter-disciplinary way.

For each project a project team is selected, grouping employees from different departments, and a Project Manager (also an employee of the company) is appointed who is entirely and exclusively responsible for the control of the project, quite independently of the company's hierarchy. The following is a brief statement of some facts and policies adopted by the company.

- Each employee works in some department.
- An employee may possess a number of skills.
- Every manager (including the MD) is an employee.
- A department may participate in none/one/many projects.
- At least one department participates in a project.
- An employee may be engaged in none/one/many projects.
- Project teams consist of at least one member.

2.4 Car Rental Co. (CRC) requires an information system whose content would include a description of cars, subcontractors (i.e. franchised garages), company expenditures, company revenues and customers. Cars are to be described by such data as: make, model, year of production, engine size, fuel type, number of passengers, registration number, purchase price, purchase date, rent price and insurance details. It is the company policy not to keep any car for a period exceeding one year.

All major repairs and maintenance are done by subcontractors CRC has long term agreements with. Therefore the data about garages to be kept in the database includes garage names, addresses, range of services and the like. Some garages require payments immediately after a repair has been made; with others CRC has made arrangements for credit facilities. Company expenditures are to be registered for all outgoings connected with purchases, repairs, maintenance, insurance, etc. Similarly the cash inflow coming from all sources - car hire, car sales, insurance claims - must be kept on file.

CRC maintains a reasonably stable client base. For this privileged category of customers special credit facilities are provided. These customers may also book in advance a particular car. These reservations can be made for any period of time up to one month. Casual customers must pay a deposit for an estimated time of rental, unless they wish to pay by a credit card. All major credit cards are accepted. Personal details (such as name, address, telephone number, driving licence number) about each customer are kept in the database.

2.5 A General Hospital consists of a number of specialized wards (such as Maternity, Paediatrics, Oncology, etc). Each ward hosts a number of patients, who were admitted on the recommendation of their own GP and confirmed by a consultant employed by the Hospital. On admission, the personal details of every patient are recorded. Separate registers need to be held to store information of the tests undertaken, diagnosis and the results of a prescribed treatment. A number of tests may be conducted for each patient. Each patient is assigned to one leading consultant but may be examined by another doctor, if required. Doctors are specialists in some branch of medicine and may be leading consultants for a number of patients, not necessarily from the same ward.

2.6 Describe a situation for which the following EAR schema might feasibly constitute an appropriate data model.

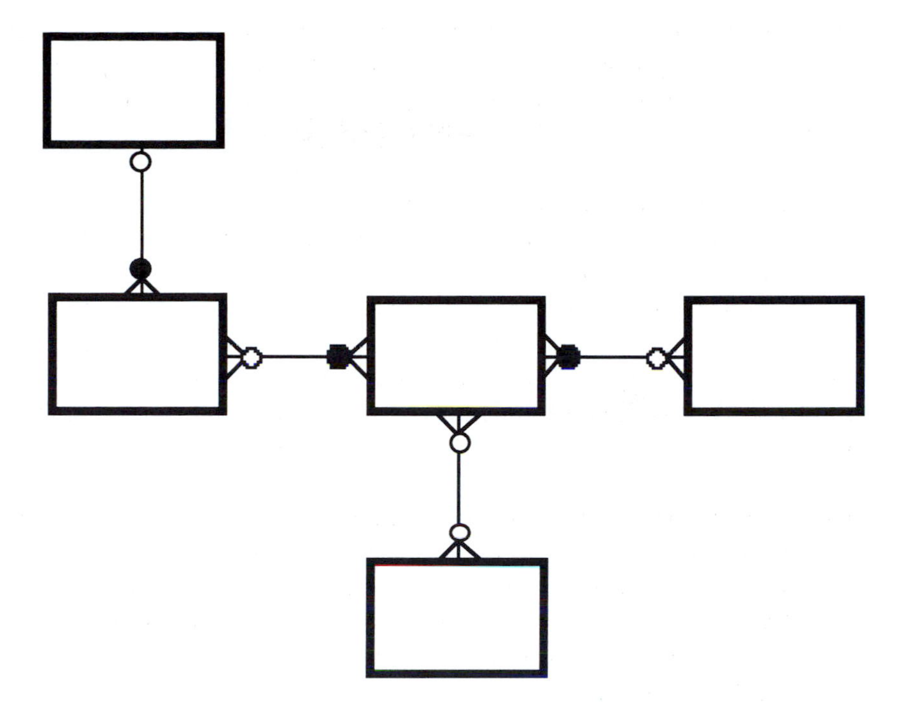

2.7 A Bus Information System is based on the representation of the street network through **nodes** (junctions, roundabouts, pedestrian crossings, etc.) and **street sections** (fragments of streets between any two nodes) to which various details held in the underlying database are (spatially) referenced. Many bus lines meet the transportation needs of the city inhabitants so they can move from virtually every location on the street network to any other by using buses (not necessarily one bus line). However, one or more bus-lines may provide a service between any two nodes. No street in the city is one-way so it is assumed that if there is a connection (*node1, node2*) then so is the opposite one (*node2, node1*). Originally it was assumed that all bus stops were all located near junctions. However, further analysis showed that this assumption was too restrictive; for longer street-sections more than one bus stop was clearly required.

To measure the safety of the city transportation system, information about traffic accidents also needs to be held in the database. Each accident being described by its location on the street section, its date and time, vehicles and people involved, etc.

CHAPTER 3

The relational model

3.1 FUNDAMENTAL CONCEPTS

In contrast with all the other models of data for database organization and management (be they hierarchical, network or, indeed, Object-Oriented), the relational model enjoys a sound theoretical foundation. In fact, providing such a foundation was one of the main objectives for the research work undertaken by E.F. Codd that resulted in the relational model being devised (Codd 1970, 1979, 1981). The model is based on three fundamental principles (Codd, 1981):

Clear distinction between the logical and physical aspects of database management (including database design, data retrieval and data manipulation) which resulted in so-called data independence.
Structural simplicity, so that all kinds of users and programmers (quite irrespective of their expertise in computing) have a common understanding of the data, and thus can communicate easily with one another about the data.
Set-oriented processing, that is the ability to express in a single statement the processing of multiple sets of records at the time.

The relational data model consists of the following three component parts:

- **structure** → a uniform single data structure type called a relation
- **manipulations** → a set of operators that transform relations into other relations
- **behaviour** → general integrity rules that guard the consistency of any database

In this chapter we shall analyse in detail the structural and the behavioural parts. The operational issues are going to be considered in Chapters 4 and 8.

The smallest unit of data in the relational model is an attribute. Attributes draw their values from suitable domains; a domain thus contains all possible values for a particular attribute. Domain definition can be done explicitly, e.g. by listing all the possible values, or by specifying conditions that all values in that domain must conform to.

Example 3.1

The possible domain definitions for the attributes *city*, *date-of-birth* and *person-name* (of people living in certain community) might be:

CITIES = {**London, Paris, New York, Tokyo**}

BIRTH-DATES = {a string of the form *dd-mm-yyyy*, such that:
01-01-1950 < *dd-mm-yyyy* < 31-12-1999
dd, mm, yyyy represent day, month year}

PERSON-NAMES = {**Smith, Jones, Green, Brown, Wilkes, Tyler**}

Note that we have made a distinction between the attributes and their corresponding domains, using small letters for the names of the former and capital letters for the latter. Note also that according to the set definition, an empty set {} is a member of any of these domains.

In computing terms, the concept of a domain can be equated to that of a (user-defined) *data type*, which perhaps makes it easier to understand the necessity for a separate treatment of domains and attributes. Looking slightly ahead, in relational theory domain definitions take some knowledge away from expressions (programs) that represent processes (transformations of relations). We are not concerned here with any specific software system but it is fair to say that, at present, no database software fully supports domains, typically allowing the users to define only very simple data types (such as numbers, characters, dates, Boolean values and similar).

The difficulty partly stems from the fact that in any given database all domains should be fully specified first - consequently the complete set of domains must be closed under every permitted operation. The other problem is that we normally allow any comparison to be made between attributes whose values are drawn from a common domain; yet the inequality **22-May-1987 < 256th-day-1989** is semantically correct despite a seemingly obvious formal (syntactic) data type mismatch. We shall consider some details involving domains in Chapter 4.

Definition 3.1

> *A relation* **R** *is a subset of an expanded cartesian product of* **n**, *not necessarily distinct domains* $\mathbf{D_1} \times \mathbf{D_2} \times \ldots, \times \mathbf{D_n}$, *such that for every element* $d^k = \langle d^k_1, d^k_2, \ldots, d^k_n \rangle \in \mathbf{R}$ *a predefined proposition* $\mathbf{p}(\langle d^k_1, d^k_2, \ldots, d^k_n \rangle)$ *is true*;
> $d^k_i \in \mathbf{D_i}$ *for every* i=1, 2,...,**n** *where* **k** *is the number of* **n**-*tuples in the relation* **R** (*cardinality of* **R**) *and* **n** *is the number of attributes in the relation* **R** (degree of **R**).

What the above definition essentially says is that a relation consists of a number of tuples, each tuple comprising a number of attributes listed in a certain order. The attributes remain in some relationship and the whole structure expresses a true proposition about the Real World. In that sense, every tuple may correspond to an entity occurrence while a relation corresponds to an entity type. This clear parallelism of EAR modelling with the relational approach is intentional here and will shortly be a matter of some consideration.

Example 3.2

Let us take a product of the domains defined in Example 3.1, that is **P** = *PERSON-NAMES* x *BIRTH-DATES* x *CITIES*. **P** will then contain all (over 76 thousand) tuples that were obtained by combining all the values from these three domains:

P = {<**Smith, 02-01-1950, London**>, <**Smith, 03-01-1950, Paris**>,
...,<**Smith, 30-12-1973, Tokyo**>,..., <**Jones, 02-01-1950, London**>,
...,<**Tyler, 02-01-1950, London**>,...,<**Tyler, 30-12-1973, Tokyo**>,
...<**Tyler, 30-12-1999, Tokyo**>}

Should the predefined proposition be 'A person with *person-name* exists, is of 40 years of age or more, was born on *date-of-birth* at *city*', the corresponding relation might only hold the following triples:

PERSON40 = {<**Smith, 02-11-1950, London**>,
<**Tyler, 02-01-1950, London**>,
<**Smith, 03-01-1950, Paris**>,
<**Jones, 02-01-1950, London**>}

PERSON40 is a subset of **P** and contains only those tuples of **P** that represent information about the 40-year old people in that community. Furthermore, the tuples with 'wrong' value for *birth-date* (that is, not the actual date of birth of the person being described), or wrong value for place of birth referred to by *cities* do not belong to PERSON40. Note also that the relation PERSON60, for example, would contain no tuples since no person born in 1950 could possibly be sixty now (May 21, 2001).

It is convenient to represent relations in a tabular form as shown in Fig. 3.1. Each row of the table represents a distinct tuple, so that its degree is a number of columns and its cardinality is a number of rows in the table.

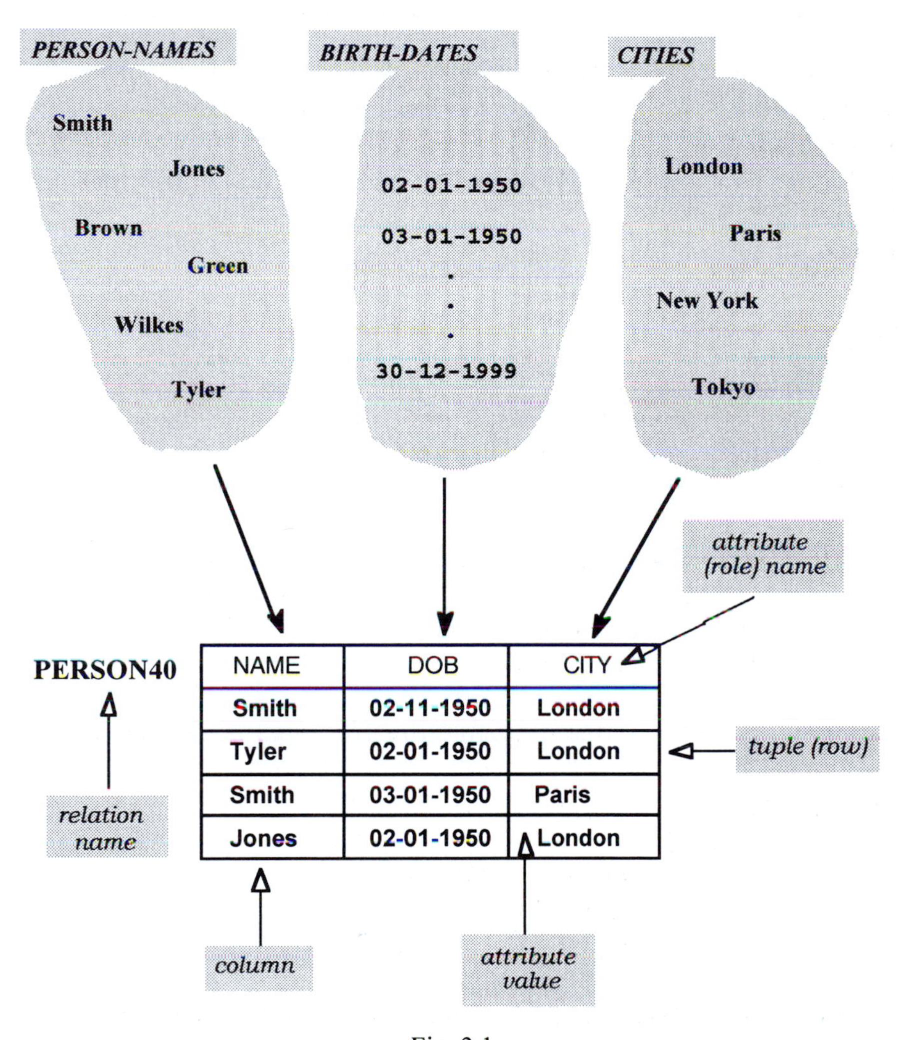

Fig. 3.1

Note that the attribute names need not be the same as the names of their underlying domains. Indeed, if two or more columns of a table are based on the same domain, they must be distinguished by being given distinct role names. From the definition, since any relation is a set, no two tuples in a relation can be identical and the order in which the tuples appear is not significant The ordering of the attributes ceases to be significant, too - as soon as the attributes are given names. The values in the relations will thus be referred to through a combination of a relation name and the attribute name.

The central concept in the relational model is that of identification of tuples solely through the attribute values. Intuitively, an attribute (or a collection of attributes) may uniquely identify any particular tuple within a relation. The existence of at least one unique key is guaranteed by definition - since there are no duplicates, the whole tuple may as well perform this role. However, the point is to have the key of **minimal** length that is composed of the smallest possible number of attributes and still retain its uniqueness.

Definition 3.2

Let **R** *be a relation defined by the proposition* $\mathbf{p}(d^k)$ *on* **n** *domains* $\mathbf{D_1} \times \ldots, \mathbf{D_n}$, *so* $d^k = <d^k_1, d^k_2, \ldots, d^k_n> \in \mathbf{R}$; *and* **R'** *be a relation defined by* $\mathbf{p'}(\delta^l)$ *on* **m** *domains* $\mathbf{D_j} \times \ldots, \times \mathbf{D_p}$, *so* $\delta^l = <d^l_j, \ldots, d^l_p> \in \mathbf{R'}$; *where* $d^k_i \in \mathbf{D_i}$ *for* $i \in \{1, 2, \ldots, \mathbf{n}\}$, $d^l_r \in \mathbf{D}_r$ *for* $r \in \{1, 2, \ldots, \mathbf{m}\}$ *and* $\mathbf{n} > \mathbf{m}$.
Then, δ *is said to be a superkey for* **R** ***if and only if*** $\mathbf{p}(d^k)$ *and* $\mathbf{p'}(\delta^l)$ *are referentially equivalent, that is whenever* $\mathbf{p}(d^k)$ *and* $\mathbf{p'}(\delta^l)$ *are both true, both* d^k *and* δ^l *refer to the same Real World entity occurrence, in which case* $k = l$ *and* $\delta \rightarrow d$ *is a bijection.*
If **m** *is minimal then* δ *is a proper candidate key for* **R.**

It is possible that a relation does have more than one **candidate key**. In such cases one of them is designated by the database designer as the **primary key**, i.e. as the primary means of identifying the corresponding tuples; all the others are **alternate keys**. The attributes participating in the primary key are termed **prime** attributes, all the others being **non-prime** attributes. A set of attributes containing a key (whether compound or single) as its subset is called a **superkey**.

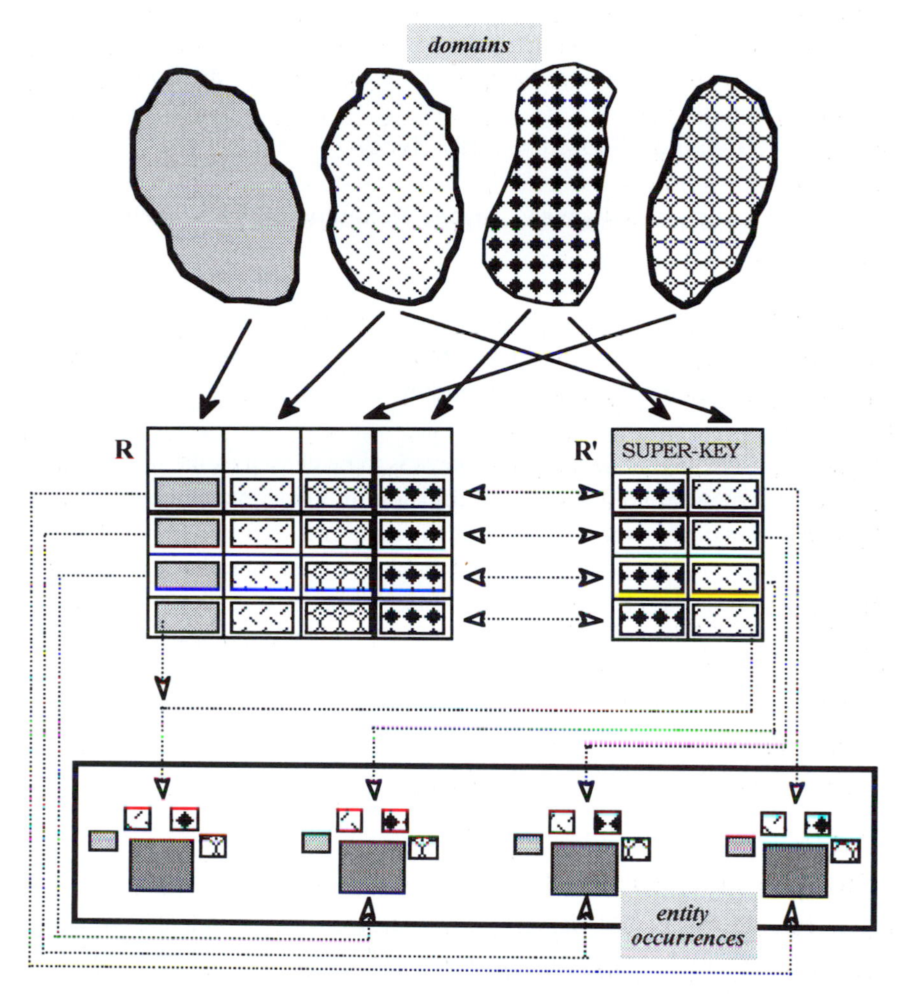

Fig. 3.2 A graphical representation of Definition 3.2

Fig. 3.2 shows the idea of Definition 3.2 albeit in a slightly simplified form. The relations R and R' are equivalent in the sense that the relevant tuples of both relations are in one-to-one correspondence to each other; furthermore every tuple from either relation refers to the particular entity occurrence in the Real World.

Clearly, keys in relations are defined by the semantics of the data - that is the essence of the predefined proposition p which associates in a certain way attributes from various domains to represent an entity occurrence in some structured form.

Example 3.3

Consider a relation that represents an entity BOOK described by the following attributes:

BOOK (Title, Author, Publisher, ISBN, Type).

Thus a tuple such as (**Ulysses, Joyce, Penguin, 0-87934-378, Fiction**) belongs to BOOK.

The possible candidate keys for BOOK are: the pair (Title, Author) and the single attribute ISBN. Any list of attributes that contains in itself either of the above keys (or both) is then a superkey.

The question of which of the candidate keys is to be chosen as the primary key is a matter of somewhat arbitrary design decision. In the above example, the most likely choice is ISBN - indeed, that was the main reason a unique book number was devised (here of course we consider a book as a concept rather than a physical copy). In general, the choice of a primary key from among the candidate keys depends on the particular circumstances, though typically the shortest key is the most favourable.

3.2 NORMALIZED RELATIONS

The relational model was devised as a formal system comprising a set of structures that hold the data (relations of various degrees) together with some operations, the basic idea being that the structures to which the predefined operations are to be applied are structurally uniform and free from any (preprocessed) substructures.

Generally speaking, a table - in the traditional understanding of this word - is not such a structure. It may contain headings and subheadings, may be structurally divided, and may contain various auxiliary results (e.g. group-sums, cross-checks, totals and subtotals) that are regarded as the *integral* elements of the table. The reasons for this become obvious when we look at such a table in the context of manual processing where recording these results facilitates and speeds up its use.

However, within the framework of the computer-processable structures, anything which is not explicitly a structural property should be eliminated from relations as it falsifies the meaning of the data and leads to undesirable update properties of these structures. Thus, the database only contains (associated somehow) pieces of raw data and not auxiliary components that are obtainable by way of processing; this can readily be done whenever required anyway.

Transforming the database relations to the acceptable (optimal) form is done by means of normalization. Normalization, being a fundamental subject of study in the relational theory, is analysed at a considerable depth later on in this book. Here we introduce, rather informally, the initial step - *First Normal Form* (1NF) - for it does not require any primitives other than those already discussed.

Example 3.4

Consider a relation that holds the information about the performances of students at a University: STUDENT (ID, Name, Record (Subject, Result))
A possible instance of this relation is depicted in Fig. 3.3. Notice that one attribute of this relation is a relation itself. The underlying domain for this attribute is itself a product of two simpler domains

SUBJECT = {**Maths, Pascal, Databases, AI, Electronics, Prolog**} and
GRADE = {**A,B,C,D,E,F**}.

A relation such as STUDENT is said to be *unnormalized* and can be converted into a *normalized* (1NF) relation as shown in Fig. 3.4.

STUDENT-RECORD

		RECORD	
ID	NAME	SUBJECT	GRADE
S011	JOHN	Mathematics	B
		Programming	A
		Databases	B
		Data Structures	C
S173	MARY	Electronics	A
		Prolog	B
		Mathematics	A
S112	PETER	Programming	C
		Prolog	A

Fig. 3.3 An unnormalized relation

ST-REC

ID	NAME	SUBJECT	GRADE
S011	JOHN	Mathematics	B
S011	JOHN	Programming	A
S011	JOHN	Databases	B
S011	JOHN	Data Structures	C
S173	MARY	Electronics	A
S173	MARY	Prolog	B
S173	MARY	Mathematics	A
S112	PETER	Programming	C
S112	PETER	Prolog	A

Fig. 3.4. A 1NF relation

The relation STUDENT-RECORD was converted into a so-called 'flat table' ST-REC by repeating the pair (*ID*, *NAME*) for every entry in the table and, consequently, removing the then superfluous role name *RECORD*. Notice that the key in this relation is a composite one (*ID*, *SUBJECT*) for *ID*, being a proper identifier for every student, it does not in itself identify tuple occurrences in ST-REC. The converted relation is in 1NF as no attribute in any of these relations is semantically decomposable, that is, all the underlying domains are simple.

Example 3.5

Consider a relation MODULES-TAKEN (St-ID, MODULE(M_1, M_2, ... M_n)). whose possible instance of this relation is depicted in Fig. 3.5. Notice that this time one attribute of this relation is a possibly variable-length list.

MODULES-TAKEN

ID	MODULES			
	M1	M2	M3	
S011	Mathematics	Programming	Databases	Data Structures
S173	Electronics	Prolog	Mathematics	
S112	Programming	Prolog	**nil**	

Fig. 3.5 An unnormalized relation

M-TAKEN

ID	MODULE
S011	Mathematics
S011	Programming
S011	Databases
S011	Data Structures
S173	Electronics
S173	Prolog
S173	Mathematics
S112	Programming
S112	Prolog

Fig. 3.6 A normalized relation

Fig. 3.6 shows the converted relation that contains the same information as the original one. The difficulty with the relation MODULE-TAKEN is that a variable-length list cannot be expressed in terms of a finite cartesian product and hence no subset of it either. Any attempt to define such a structure (in whatever programming system) must rely either on an assumed value for the maximum number of attributes (always uncertain) or on the properties of dynamic data structures such as pointer-based linked list.

The first solution will, in almost all cases, lead to a situation where either some attribute values cannot be recorded at all (as demonstrated in Fig. 3.5) or many fields will hold null values - or both! The second idea is a good one but not within the framework of the relational model since no dynamic structure exists there and all associations between the tables are represented through attribute values rather than pointers or links.

Concluding the above examples, a relation is in 1NF if every attribute is defined on a simple domain. It is a rather rudimentary operation to convert any relation to its normalized form - it is an entirely syntactic operation not involving any deep meaning of the data, nor do the relationships play any role in it.

The relational theory deals almost exclusively with normalized relations (save for object-relational structures described in Chapter 9). The reason for this restriction is that the primitive relational operations (such as project, join, select, etc. - see Chapter 4) that form a basis for expressing retrieval and updating processes are much simpler and there are fewer of them.

3.3 INTEGRITY CONSTRAINTS

A shared multiuser database can contain millions of records of data that represent some required fragment of the Real World. For the owner of all that information (be it an organization, a company, an enterprise, etc.) the database is then a valuable asset - information held there must be true and consistent as this information itself is a company asset, just as the other company resources (workforce, capital, holdings, machines, etc.) are.

Numerous conditions may guard the consistency of data in a database. For instance, the values held in domains can be restricted by specifying some conditions as to the form or meaning (or both). Examples are: numerical values restricted to positive rational numbers with two decimal places, parity conditions for numerals, self-detecting codes for certain strings of literals, date values restricted in time and similar. Such constraints are said to be **particular**, that is specific for the application system being devised. This type of integrity constraint essentially defines the objects and their representation in the system.

Relational theory also provides two **general** integrity constraints - the constraints that apply to every database, irrespective of its content or area of applicability.

Definition 3.3 **Entity Integrity**

No prime attribute of a relation may hold a null value.

It follows from Definition 3.2 that a primary key $\delta = \langle d_j, \ldots, d_p \rangle \in \mathbf{R}$ with a predefined true proposition $\mathbf{p}(\delta)$ is unique, minimal and refers to some specific Real World entity occurrence. Now, suppose some attribute of an occurrence δ is null thus $\delta^l = \langle d^l_j, \ldots .\mathbf{NULL}, .., d^l_p \rangle$. In such a case the truth-value of $\mathbf{p}(\delta^l)$ is undefined - since the number of attributes in δ is minimal and one argument of the proposition $\mathbf{p}$ is missing, the value of $\mathbf{p}(\delta^l)$ cannot be found. Also, the occurrence δ^l loses its uniqueness and does not refer to any Real World entity.

Example 3.6

Let DIRECTORY (FNAME, SNAME, ADDRESS, TELEPHONE) represent a telephone directory of a certain area. Suppose, for the sake of this example, that the key in this relation is (*FNAME, SNAME*). An instance of DIRECTORY is depicted in Fig. 3.7.

DIRECTORY

FNAME	SNAME	ADDRESS	TELEPHONE	
John	Adams	1 High Str	01234 123456	
Mary	Adams	11 North Av	01234 234567	
	Adams	9 South Av	01234 345678	not allowed
Peter		5 Dore Clo	01234 456789	not allowed
		13 Back Str	01234 567890	not allowed

Fig. 3.7 A fragment of telephone DIRECTORY

Neither of the three tuples marked *not allowed* represent any useful information to a potential user, nor do they refer to any Real World entity occurrence.

Example 3.7

Let CREDIT (ACCOUNT, NAME, LIMIT) with a primary key ACCOUNT be a relation in a bank database. Fig. 3.8 shows its possible instance.

CREDIT

ACCOUNT	NAME	LIMIT	
12345678	Smith	1250	
23456789	Jones	3000	
34567890	Jones	5100	
	Brown	2800	not allowed
56789012	Green	900	

Fig. 3.8 A fragment of the relation ACCOUNT

The reasons for not permitting NULL-values to be held by the primary key are exposed in this simple example quite well. Partial information that some (undefined!) account enjoys some credit limit does not present any true statement about the customer concerned (who may actually have a number of other accounts) nor does it permit proper calculations or linking with the other tables maintained in the database.

Definition 3.4 **Foreign Key**

Let **R** *be a relation* (Δ, X) *with a primary key* Δ, *and* **S** *be a relation* (A, Γ) *where* Δ *and* Γ *are defined on the same not necessarily simple domain. Then,* Γ *is said to be a foreign key in* **S**.

This definition essentially says that an attribute (or a collection of attributes) in a relation may perform the role of a primary key in another relation. Notice that there is no requirement for a foreign key to be a component of the primary key (or otherwise) of its 'home' relation and it need not have the same role name as the corresponding key in the other relation.

Definition 3.5 **Referential Integrity**

If a relation **S** *includes a foreign key* Γ *matching the primary key* Δ *of some other relation* **R** *then every value of* Γ *in* **S** *must be equal to the value of* Δ in some tuple of **R** *or be wholly null.*

Example 3.8

Let COURSE (MODULE, LECTURER-ID) with the primary key MODULE and REGISTER (S_ID, SUBJECT, GRADE) with the primary key (S_ID, SUBJECT) be relations in a University database.

Fig. 3.9 shows possible instances of both relations. The first one describes all the modules available for the students in the University while the second holds information on students' results. In the relation REGISTER, the attribute SUBJECT is a foreign key - its values must correspond to those of MODULE from COURSE, otherwise some students might have been awarded grades on the subjects the University does not provide!

The notions of primary and foreign keys are fundamental to the relational theory. The primary keys provide the only means of logical addressing of tuples in a relation. Recall that a relation being a set of tuples does not impose any ordering on them. There is no explicit (physical) linking of tuples between relations - all relationships between data are represented solely by the attribute values! For the sake of presentation the results of some actions (e.g. retrievals) on the database tables can be printed or displayed in some specific order (for instance sorted alphabetically by the employee name) but this - or any other kind of ordering - shall not be used in any legal database operations.

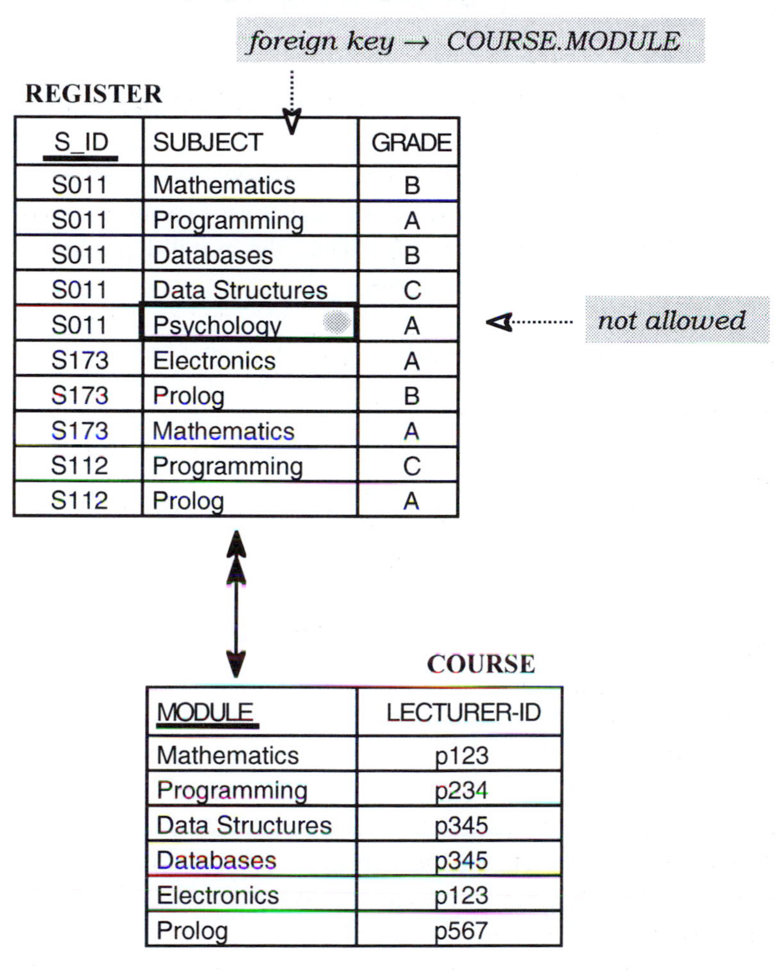

REGISTER

S_ID	SUBJECT	GRADE
S011	Mathematics	B
S011	Programming	A
S011	Databases	B
S011	Data Structures	C
S011	Psychology	A
S173	Electronics	A
S173	Prolog	B
S173	Mathematics	A
S112	Programming	C
S112	Prolog	A

COURSE

MODULE	LECTURER-ID
Mathematics	p123
Programming	p234
Data Structures	p345
Databases	p345
Electronics	p123
Prolog	p567

Fig. 3.9

Neither the necessity of the primary key nor the principle of entity integrity imply that the only way to access the data is via the primary key. Indeed, as it will become clear in Chapter 4, the data can be accessed through **any** combination of attributes whatsoever - though in such accesses a set of tuples (rather than a single one) is likely to be retrieved.

In the absence of any other structure but relation, referential integrity ensures that cross-relationships between tuples in different relations are maintained. This aspect of relational databases is specifically important in the context of storage operations, that is operations that change the content or the structure of a database. This issue will be dealt with in some detail in Chapter 6.

3.4 REPRESENTATION OF EAR MODELS BY RELATIONS

The Real World (or more precisely that part of it which is to be implemented as a database) is perceived by the developers through a set of entities described by attributes and interrelated by relationships. Thus, the overall logical structure of the database is expressed by EAR diagrams.

On the other hand, the relational database can represent a somehow abstracted user's view of the relevant part of the Real World. The relational model - as so far defined - consists though of one type of well-defined object, that is normalized relations restricted further by the entity integrity and referential integrity constraints. The fundamental feature of the relational model is that all associations between the tuples of different relations are represented solely by the data values of the attributes defined on common domains.

Clearly, we need a transition mechanism that would unambiguously translate the results of EAR modelling into a set of appropriate relations. Loosely speaking, both the entities and relationships among them get represented by relations. More precise definitions and examples follow.

Definition 3.6

Let **E** *be an entity type with attributes* $\{e^k_1, e^k_2, ..., e^k_n\}$ *such that* $e^k_i \in \mathbf{D}_i$ *for every* $i = 1, 2,..., n$.
A relation $\mathbf{R} = \{r^k : r^k = \langle r^k_1, r^k_2, ..., r^k_n \rangle, r^k_i \in \mathbf{D}_i, i = 1, 2,...,n\}$ *is said to represent the entity* **E** *if* $r^k_i = e^k_i$.

The above definition states that an entity gets represented by a table with **n** distinct columns, each of which corresponds to one of the **n** attributes of that entity; every tuple (row in the table) corresponds in turn to one of the **k** entity occurrences.

Definition 3.7

> *Let **R** be a relationship involving entity types $\mathbf{E}_1$ and $\mathbf{E}_2$. Let $\mathbf{R}_1$ with the primary key δ^k and $\mathbf{R}_2$ with the primary key η^j be the relations representing $\mathbf{E}_1$ and $\mathbf{E}_2$ respectively. Then the relation $\mathbf{Q} = \{<\delta^k, \eta^j>\}$ is said to represent the relationship **R**.*

Thus, a relationship between two entities is represented by a table **Q** composed of the two primary keys of the relations involved. The number of tuples in the relation **Q** depends on the type of the relationship. For instance, a *one-to-one* two-sided compulsory relationship yields k=j tuples in **Q**, a *one-to-many* two-sided compulsory relationship gives j tuples in **Q**, a *many-to-many* relationship may result in maximum k x j tuples in **Q** (this would happen if exactly every occurrence of $\mathbf{E}_1$ is associated with exactly every occurrence of $\mathbf{E}_2$).

Case Study

Here we are going to devise a relational representation of a part of a University database limiting our considerations to the following five entities STUDENT, REGISTER, MODULE, LECTURER and OFFICE. Fig. 3.10.a depicts a relational representation of the two entities OFFICE and LECTURER (primary keys are underlined). Since the relationship between these entities is *one-to-one*, all the attributes from either relation can simply be posted to the remaining one, thus avoiding a superfluous table for this relationship. This is shown in Fig. 3.10.b and 3.10.c.

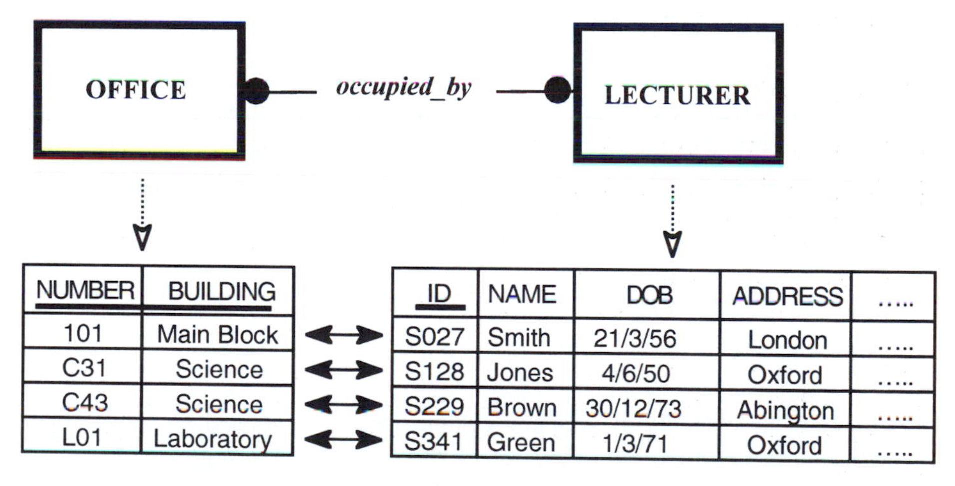

NUMBER	BUILDING
101	Main Block
C31	Science
C43	Science
L01	Laboratory

ID	NAME	DOB	ADDRESS	.....
S027	Smith	21/3/56	London	.....
S128	Jones	4/6/50	Oxford	.....
S229	Brown	30/12/73	Abington	.....
S341	Green	1/3/71	Oxford	.....

Fig. 3.10.a

ID	NAME	DOB	ADDRESS	.....	NUMBER	BUILDING	
S027	Smith	21/3/56	London	.....	101	Main Block	
S128	Jones	4/6/50	Oxford	.....	C31	Science	
S229	Brown	30/12/73	Abington	.....	C43	Science	
S341	Green	1/3/71	Oxford	.....	L01	Laboratory	
					N101	New Block	**X**
S678	Green	10/4/75	Oxford				**OK**

Fig. 3.10.b

NUMBER	BUILDING	ID	NAME	DOB	ADDRESS	.....	
101	Main Block	S027	Smith	21/3/56	London	.....	
C31	Science	S128	Jones	4/6/50	Oxford	.....	
C43	Science	S229	Brown	30/12/73	Abington	.....	
L01	Laboratory	S341	Green	1/3/71	Oxford	.....	
N101	New Block						**OK**
		S678	Green	10/4/75	Oxford	.....	**X**

Fig. 3.10.c

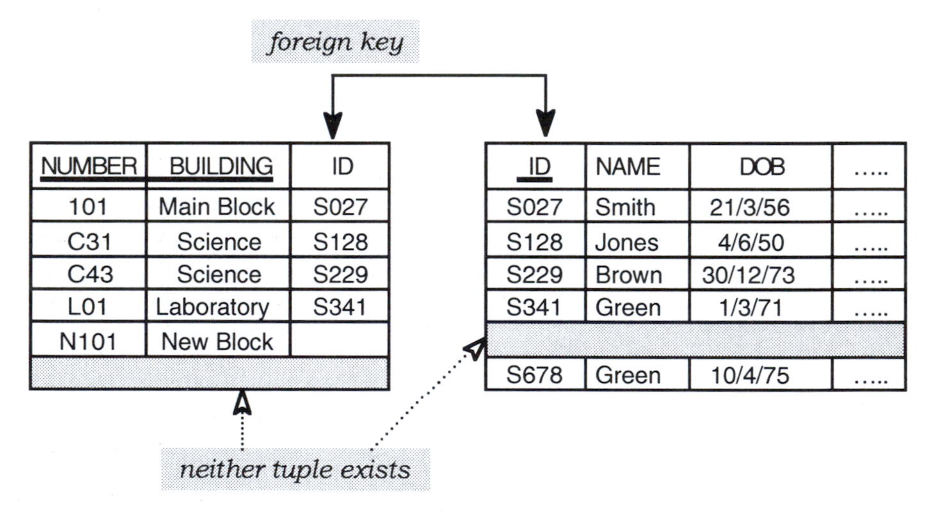

Fig. 3.10.d

However, if the relationship were non-compulsory, both tables depicted in Figs. 3.10.b, c may exhibit certain kinds of misbehaviour. For instance, we cannot register the fact that some offices stay empty since the attribute forming the primary key (ID) would have to accept a NULL value which is contradictory to the rule of entity integrity (NULL values are allowed though for the attributes NUMBER and BUILDING since neither of these is part of a primary key).

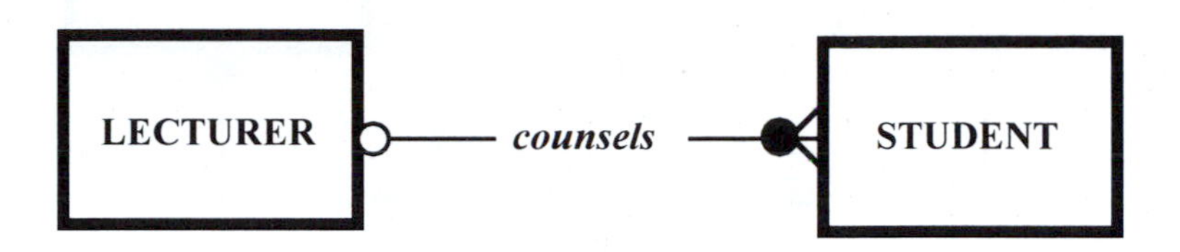

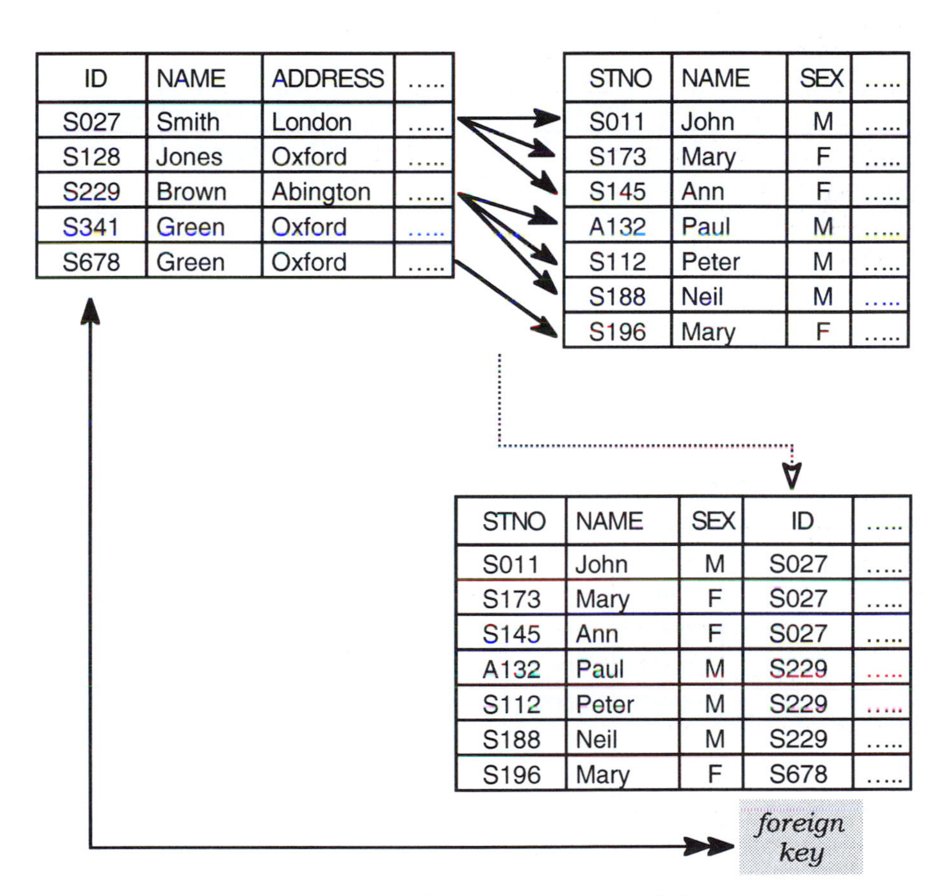

ID	NAME	ADDRESS	.....
S027	Smith	London	.....
S128	Jones	Oxford	.....
S229	Brown	Abington	.....
S341	Green	Oxford	.....
S678	Green	Oxford	.....

STNO	NAME	SEX	.....
S011	John	M	.....
S173	Mary	F	.....
S145	Ann	F	.....
A132	Paul	M	.....
S112	Peter	M	.....
S188	Neil	M	.....
S196	Mary	F	.....

STNO	NAME	SEX	ID	.....
S011	John	M	S027	.....
S173	Mary	F	S027	.....
S145	Ann	F	S027	.....
A132	Paul	M	S229	.....
S112	Peter	M	S229	.....
S188	Neil	M	S229	.....
S196	Mary	F	S678	.....

Fig. 3.11 Representation of *one-to-many* relationship

In the second case, we would not be able to register the fact that actually some lecturers may not occupy any office, since neither of (NUMBER, BUILDING)

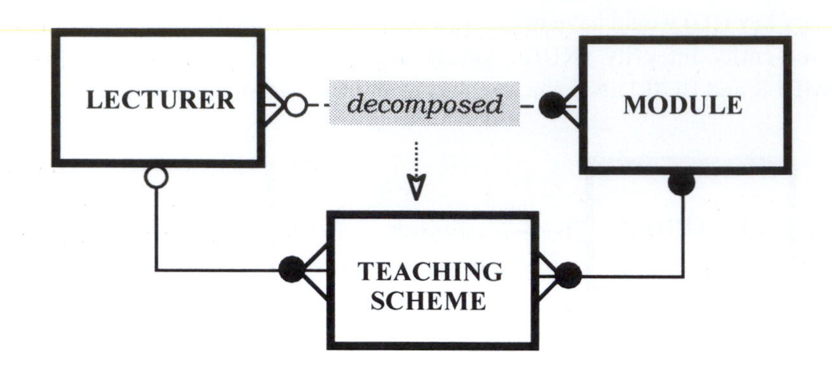

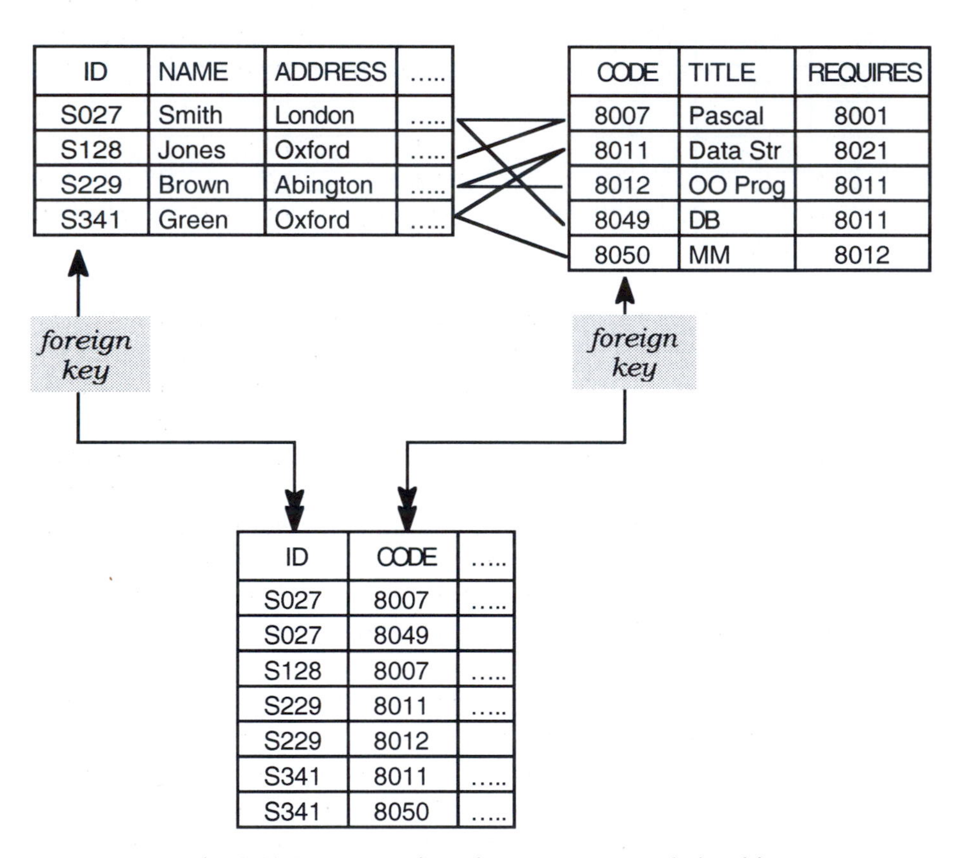

ID	NAME	ADDRESS	.....
S027	Smith	London	.....
S128	Jones	Oxford	.....
S229	Brown	Abington	.....
S341	Green	Oxford	.....

CODE	TITLE	REQUIRES
8007	Pascal	8001
8011	Data Str	8021
8012	OO Prog	8011
8049	DB	8011
8050	MM	8012

ID	CODE	.....
S027	8007	.....
S027	8049	
S128	8007	.....
S229	8011	.....
S229	8012	
S341	8011	.....
S341	8050	.....

Fig. 3.12 Representation of *many-to-many* relationship

could accept a NULL value due to, again, the rule of entity integrity. The optimal solution is shown in Fig. 3.10.d, where the entity OFFICE and the linking relationship are both represented by one table which was achieved by duplicating LECTURER.ID and posting it to the relation OFFICE thus adding a referential integrity connection.

Let us now consider a *one-to-many* relationship that involves the entities LECTURER and STUDENT (see Fig. 3.11). Here again, the associations between various tuples of the two tables can be recorded by duplicating the attribute LECTURER.ID and posting it to the relation STUDENT (since no student has more than one Personal Tutor).

The *many-to-many* relationship that links the entities LECTURER and COURSE gets represented by a table TEACHING-SCHEME formed by the respective primary keys, i.e. LECTURER.ID and MODULE.CODE and hence the associations between the relevant tuples are recorded explicitly (see Fig. 3.12). Both attributes are needed to form the primary key in TEACHING-SCHEME and each of them is a foreign key to match its values with a corresponding primary key in LECTURER and COURSE, respectively. This way of representation agrees, of course, with the principle of decomposing *many-to-many* relationships as described in the previous chapter.

The convoluted relationship such as *'a module is a pre-requisite for another module'* essentially introduces certain logical ordering (such as tree or graph) in the relation and hence an *intra-referential* integrity constraint - one attribute references another within the same relation. This means that whenever a new module is to be inserted, the insertion mechanism must ensure that the module's pre-requisite is already represented at its home 'position' in the relation; otherwise an insertion of a module with a non-existent prerequisite would have been attempted (see Fig. 3.13). Similarly, a deletion of a module may require further action, that is updating of all relevant tuples where the module to be deleted existed as pre-requisite for some other modules.

A possible relational representation of models relying on the concept of super/sub-entities is shown in Fig. 3.14. We can observe that some amount of redundancy is there - the primary key is replicated in all three tables to ensure both identification and referential correspondence between the relevant tuples. As before, the mechanisms for insertion and deletion must ensure that an action on the 'super' table automatically invokes the appropriate action on the relevant 'sub' table.

In general, the EAR models based on super/sub-entities (essentially the concept of the so called *is_a* relationship) do not easily lend themselves to relational representation; rather they form part of Object-Oriented databases. We will return to this in Chapter 9.

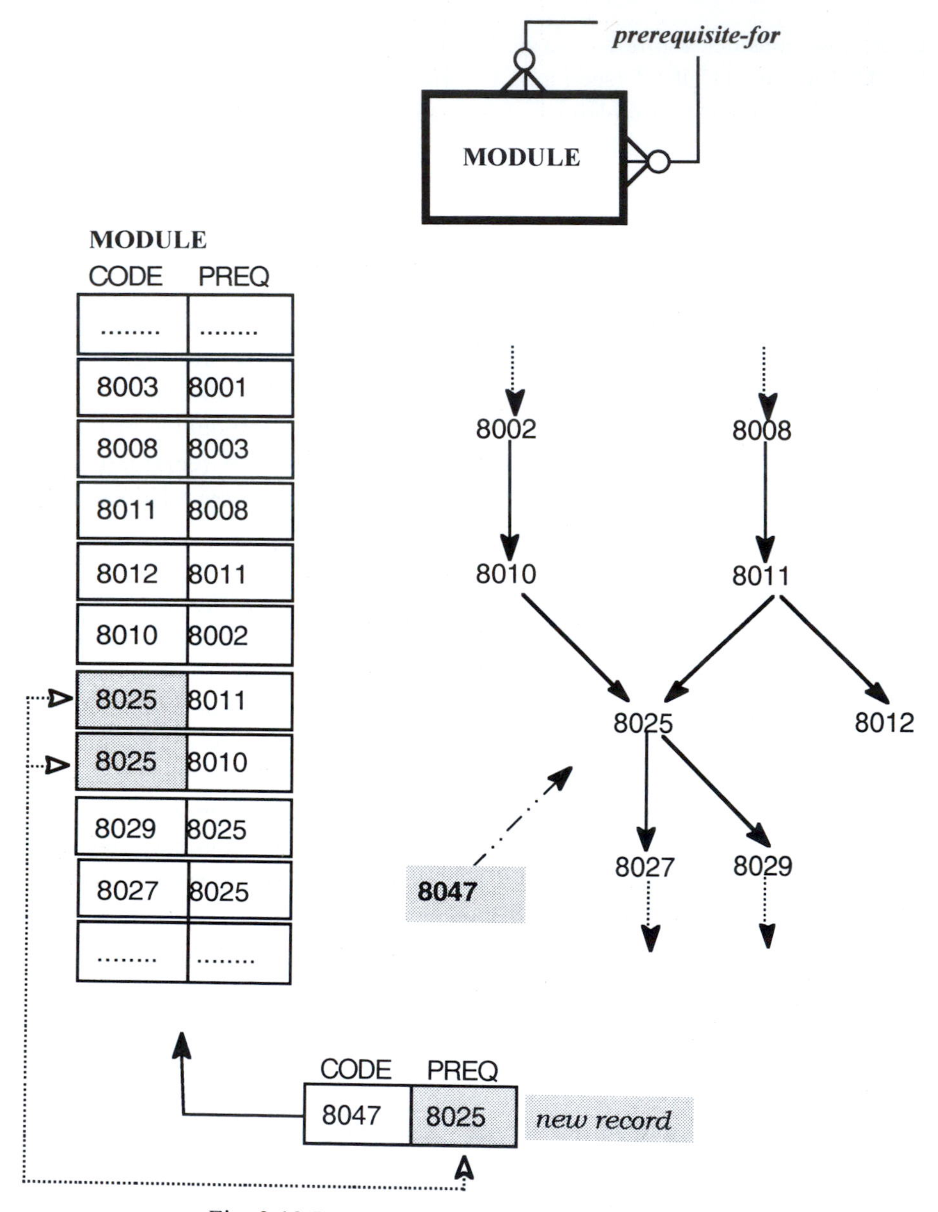

Fig. 3.13 Representation of *convoluted* relationships

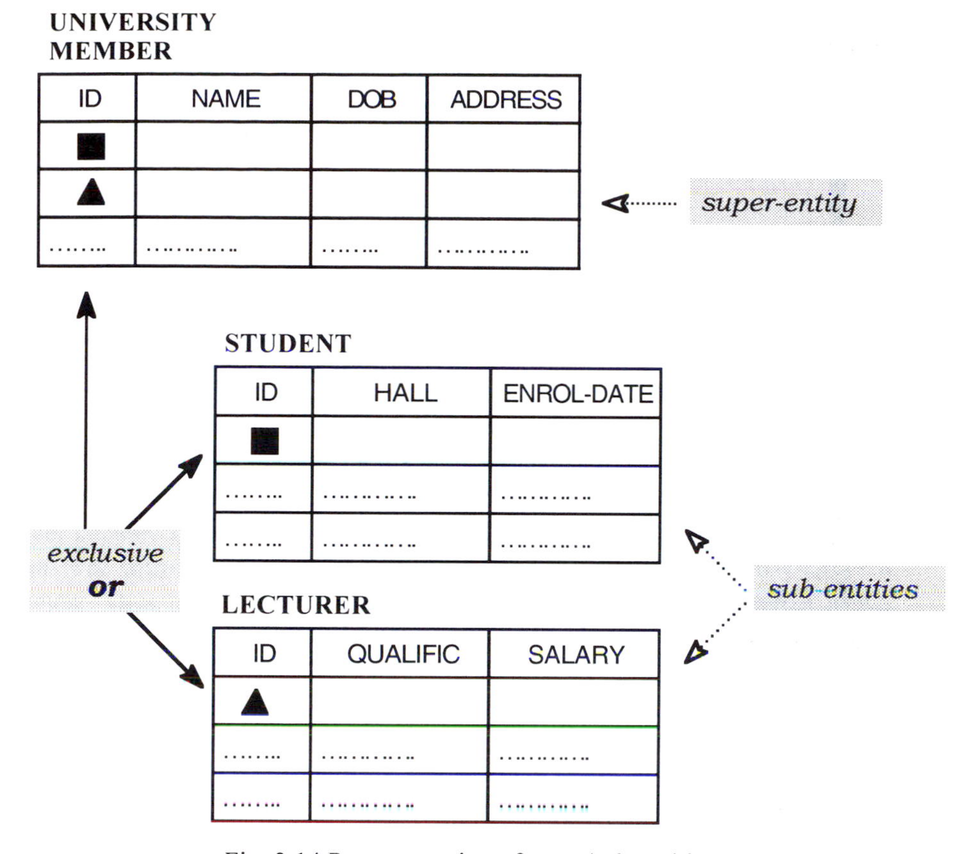

Fig. 3.14 Representation of super/sub-entities

The final effect of representing the logical (EAR) data model is shown in Fig. 3.15 in the form of a referential integrity diagram for (a fragment of) the University Database. It constitutes our statement, presented in a transparent yet precise manner, of our perception of the relevant part of the computerizable conceptual data model expressed in relational terms.

not held in **S**, again subject to the condition of **R** and **S** being *union compatible.* As previously, there is no requirement for the relevant attribute names to be the same.

Definition 4.3 **Intersection**

> *Let* $\mathbf{R} = \{\mathbf{r}^k = <r^k_1, r^k_2, .., r^k_n>\}$ *and* $\mathbf{S} = \{\mathbf{s}^i = <s^i_1, s^i_2, .., s^i_m>\}$.
> *Then a relation* $\mathbf{Q}=\mathbf{R} \cap \mathbf{S} = \{\mathbf{r}^k\} \cap \{\mathbf{s}^i\}$ *is said to be an intersection of* **R** *and* **S** *provided that* **R** *and* **S** *are union compatible.*

Intersection produces a relation that contains all those tuples of **R** that are also tuples of **S** (i.e. the common part of both relations). The condition of union compatibility is again in force.

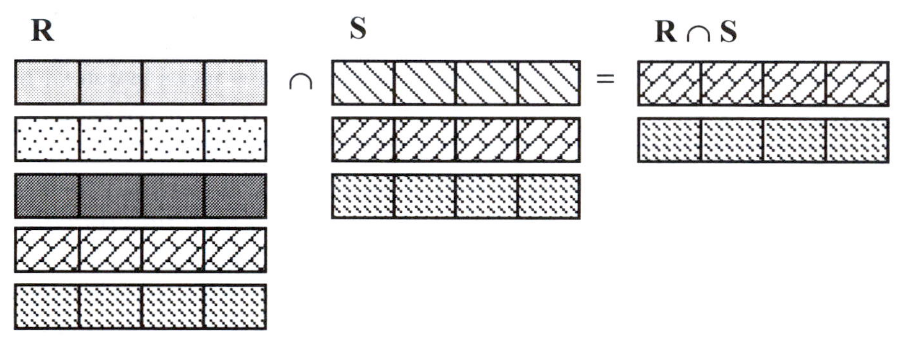

Fig. 4.5 Schematic representation of *intersection*

Definition 4.4 **Product**

> *Let* $\mathbf{R} = \{\mathbf{r}^k = <r^k_1, r^k_2, ..., r^k_n>\}$ *and* $\mathbf{S} = \{\mathbf{s}^i = <s^i_1, s^i_2, ..., s^i_m>\}$.
> *Then a relation* **Q** *that contains all combinations of concatenated pairs* **r** *and* **s**
>
> $$\mathbf{Q} = \mathbf{R} \otimes \mathbf{S} = \{\mathbf{r}^m \sim \mathbf{s}^n : \text{m=1, 2,...,k } \textit{and} \text{ n=1, 2,...,i}\}$$
>
> *is said to be a product of* **R** *and* **S** *where the concatenation operator* $\sim$ *is defined as:* $\mathbf{r}^k \sim \mathbf{s}^i = r^k_1, r^k_2, ..., r^k_n, s^i_1, s^i_2, ..., s^i_m$.

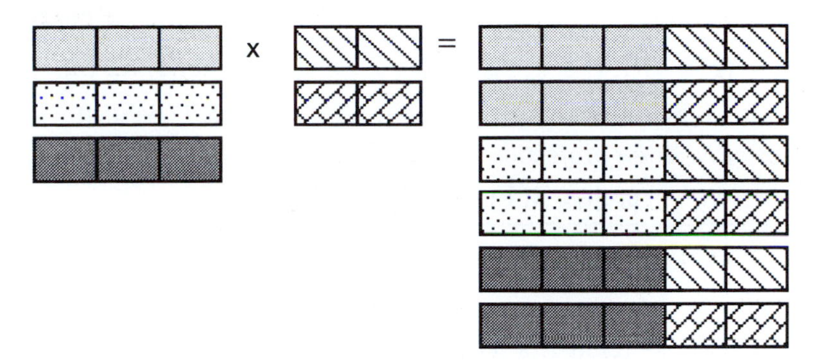

Fig. 4.6 Schematic representation of *product*

The Cartesian *product* is a binary operation that combines two relations together yielding a new relation whose tuples contain all the attributes originally held by the tuples of **R** and **S**. Notice that the resulting relation may hold duplicate attributes if some attributes of **R** and **S** are defined on common domains. Note also that the condition of union compatibility does not apply here. In theory, *product* can be applied to any two relations but in practice it is hardly used at all. On its own, it carries little semantic value, that is very rarely could it represent any meaningful query one would wish to pose to a database. Besides, product causes some adverse quantitive effects. For instance, if both R and S are of size 1K their product will result with a relation of 1M!

So much for the operators that were adopted from set theory. We will shortly demonstrate their use on a number of examples but, to make these examples more comprehensible, the other operators that were specifically devised to facilitate the effective use of the relational model, need to be described first. For some of these operations, a concept of so-called *comparability* will be needed. Intuitively, whenever two attributes are to be related to each other by comparators such as $=$, $\neq$, $>$, $\geq$, $<$, $\leq$ etc., the comparison must be meaningful. For instance, we can compare two attributes whose values are dates (say), but relating in this sense *birth-date* to *account-balance* is meaningless.

Definition 4.5

Attributes X *and* Y *are said to be* ***Θ–comparable*** *if* X Θ Y *is either true or false but not undefined, and* Θ *is any comparator from a set of allowable ones within the system.*

Definition 4.6 **Projection**

> *Let* $\mathbf{R} = \{\mathbf{r}^i = <r^i_1, r^i_2, ..., r^i_n>\}$. *Let X denote the set of all the attributes of* $\mathbf{R}$, thus $\mathbf{r}^j(X) = <r^j_1, r^j_2, ..., r^j_n>$, *and* T *be any subset of* X. *Then projection of* $\mathbf{R}$ *on* T *is defined as:*
>
> $$\textbf{project } \mathbf{R}(T) = \{\mathbf{r}^k(T):\ r \in \mathbf{R}\},\ k \le i$$

Projection then is a unary operation that reduces the operand to the attributes indicated; since the result is a relation all duplicate tuples are discarded. The meaning of this operation could be expressed by a query: *In a given relation, what are the different values of the specified collection of attributes?*

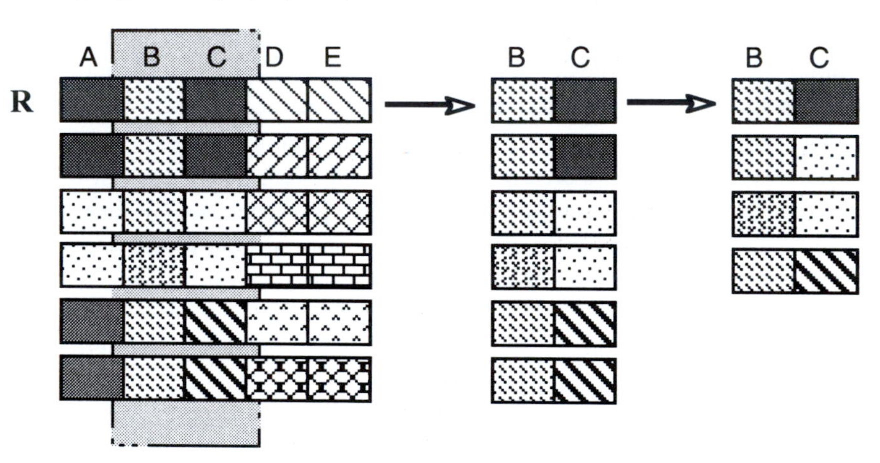

Fig. 4.7. Schematic representation of *projection*

Example 4.1

Let **BOOK** (<u>REF#</u>, AUTHOR, TITLE), **INDEX** (<u>AUTHOR, TITLE</u>, CLASS#, SHELF#) and **SUBJECT** (<u>CLASS#</u>, NAME) be relations that describe books held in a library (primary keys underlined). The relation **BOOK** holds some details of every physical copy while **INDEX** indicates the location in which all copies of the same book are held; **SUBJECT** gives the usual content-dependent classification of books. Fig. 4.8 depicts possible instances of these three relations.

BOOK

REF#	AUTHOR	TITLE
R003	Joyce	Ulysses
R004	Joyce	Ulysses
R023	Greene	Short Stories
R025	Orwell	Animal Farm
R033	Lem	Robot Tales
R034	Lem	Return from the Stars
R036	Golding	Lord of the Flies
R028	King	Strength to Love
R143	Hemingway	Death in the Afternoon
R149	Hemingway	To Have and have Not

SUBJECT

CLASS#	NAME
c1	Fiction
c2	Science-Fiction
c3	Non-Fiction
c4	Science
c5	Poetry
c6	Drama

INDEX

AUTHOR	TITLE	CLASS#	SHELF#
Joyce	Ulysses	c1	12
Greene	Short Stories	c1	14
Orwell	Animal Farm	c1	12
Lem	Robot Tales	c2	23
Lem	Return from the Stars	c2	23
Golding	Lord of the Flies	c1	12
King	Strength to Love	c3	24
Hemingway	Death in the Afternoon	c3	22
Hemingway	To Have and Have Not	c1	12

Fig. 4.8 A library database

The result of projecting **BOOK** over the attribute AUTHOR is shown in Fig. 4.9; it is an answer to the query: *Who are the authors of the books kept in the library?*

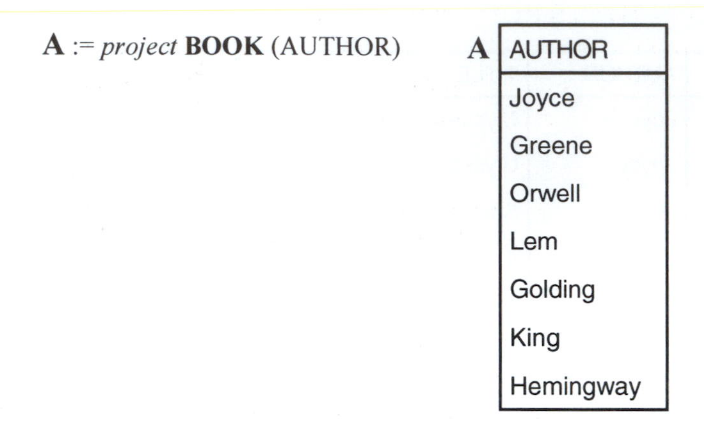

A := *project* **BOOK** (AUTHOR)

A

AUTHOR
Joyce
Greene
Orwell
Lem
Golding
King
Hemingway

Fig. 4.9

As already mentioned the relational operators can be combined and used repeatedly; thus for instance the query *What subjects are not represented in the library at all?* can be expressed as:

project **SUBJECT** (CLASS#) \ *project* **INDEX** (CLASS#)

and its logical mechanism of execution is presented in Fig. 4.10.

project **SUBJECT** (CLASS#) \ *project* **INDEX** (CLASS#)

CLASS#
c1
c2
c3
c4
c5
c6

\

CLASS#
c1
c2
c3

→

CLASS#
c4
c5
c6

Fig. 4.10

The first projection **SUBJECT**(Class#) produces a unary relation containing all allowable class numbers, the second projection **INDEX**(Class#) yields the actual class numbers. The difference of these two relations is then a table that comprises the class numbers not referred to in the table of all books held in the library. Note that these two intermediate relations are necessarily *union-compatible* as holding a single attribute defined on the same domain.

Definition 4.7 **Restriction**

> *Let* $\mathbf{R} = \{\mathbf{r}^i = <r^i_1, r^i_2, ..., r^i_n>\}$. *Then* Θ-*restriction of* **R** *on attributes* r_p *and* r_q *is defined as:*
>
> $$\textbf{restrict } \mathbf{R}(r^i_p \ \Theta \ r^i_q) = \{\mathbf{r}^k\text{: } r^k_p \ \Theta \ r^k_q = \text{true } \textbf{and } \mathbf{r}^k \in \mathbf{R}\}, \quad k \in \{1..i\};\ p, q \in \{1..n\}$$
>
> *provided* p^{th} *attribute is* **Θ-comparable** *with* q^{th} *attribute of* **R**.

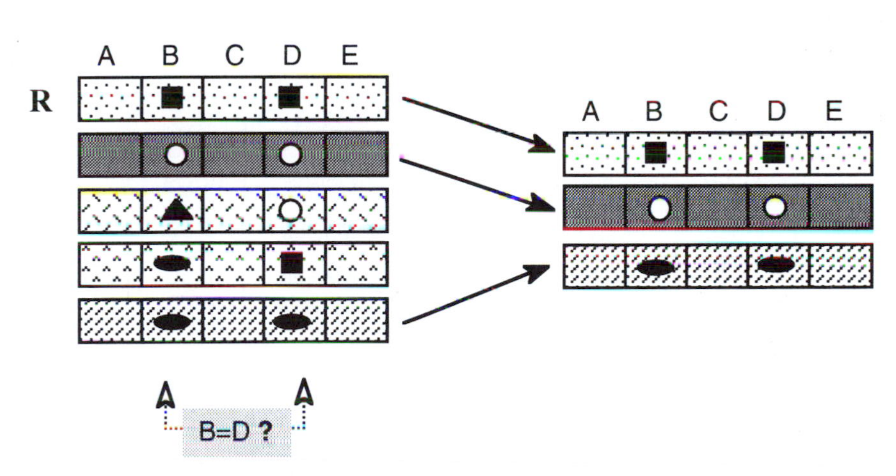

Fig. 4.11 Schematic representation of *restriction*

Restriction eliminates from a given relation all those tuples that do not conform to the condition specified in the restriction clause; note that the attributes involved in the condition must necessarily be **Θ-comparable**. The restriction clause can be built up from a number of simple conditions connected by the logical operators {**and**, **or**, **not**} but none of these simple conditions contain any explicit reference to any attribute value.

Example 4.2

Occasionally the library may sell some books at auction and all the transactions are registered in the relation **AUCTION**. The operation

restrict **AUCTION** (PRICE > COST)

will produce a relation that contains all those books sold for more than originally paid for.

AUCTION

REF#	BOUGHT	COST	SOLD	PRICE
R005	17-Mar-84	25.00	23-Sep-86	12.50
R020	02-Dec-43	4.00	17-Oct-88	145.50
R022	09-Nov-79	7.50	21-Nov-88	3.25
R048	15-May-68	3.50	16-Mar-89	8.50
R049	15-May-68	3.50	16-Mar-89	8.50
R073	27-Aug-76	18.50	25-Mar-89	9.25

restrict **AUCTION** (PRICE > COST)

REF#	BOUGHT	COST	SOLD	PRICE
R020	02-Dec-43	4.00	17-Oct-88	145.50
R048	15-May-68	3.50	16-Mar-89	8.50
R049	15-May-68	3.50	16-Mar-89	8.50

Fig. 4.12

Definition 4.8 Selection

Let $\mathbf{R} = \{\mathbf{r}^i = <r^i_1, r^i_2, ..., r^i_n>\}$. *Then selection from* **R** *subject to condition* $\alpha(r^i_p, r^i_{p+1}, ..., r^i_q)$ *is defined as:*

$\textbf{select } \mathbf{R}(\alpha) = \{\mathbf{r}^k\text{: } \alpha(r^k_p, r^k_{p+1}, ..., r^k_q) = \text{true} \textbf{ and } \mathbf{r}^k \in \mathbf{R}\},$
$k \in \{1..i\};\ p,\ q \in \{1..n\}$

$\alpha(r_p, r_{p+1}, ..., r_q)$ *is a predicate composed of elements* $(r_m \Theta X)$ *interconnected by the logical operators* {**and**, **or**, **not**}, X *is a value acceptable for the attribute* r_m *and* $m \in \{p...q\}$

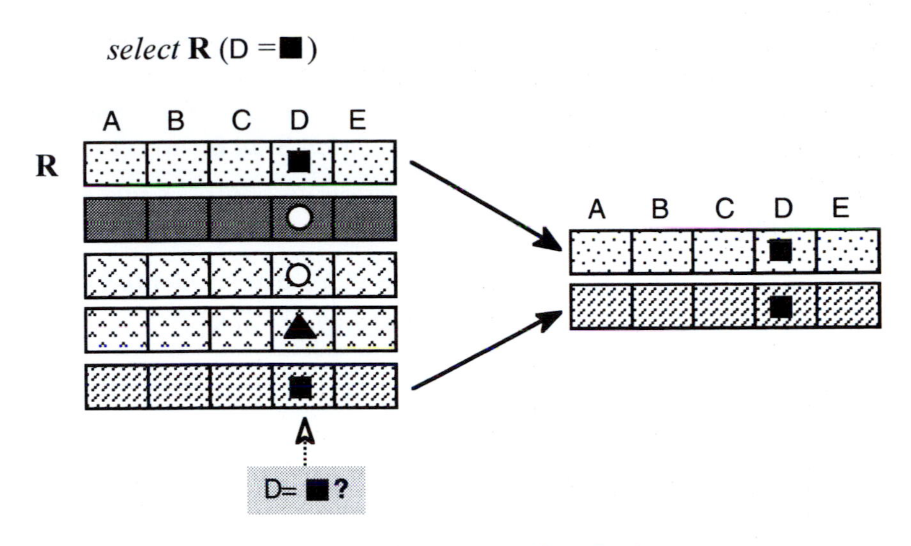

Fig. 4.13 Schematic representation of *selection*

Selection - similar *to restriction* - eliminates from a given relation all those tuples that do not conform to a specified condition. Note that the condition here is expressed in a slightly different way; being composed of elements of the form (*attribute name* **Θ** *attribute value*) it involves values explicitly. Quite frequently, however, these two operations (though different from a semantic viewpoint) are not being distinguished at all since the following property holds:

select (**restrict R**(β))(α) = **restrict** (**select R**(α))(β)

for all valid conditions α and β.

Example 4.3

A typical query of a customer in the library aims to obtain information concerning a location of a non-uniquely specified book e.g. *Find a non-fiction book by Hemingway*. This query, expressed in terms of selection as:

select **INDEX** (AUTHOR = 'Hemingway' and CLASS# = 'c3')

will produce the relation as shown below

AUTHOR	TITLE	CLASS#	SHELF#
Hemingway	Death in the Afternoon	c3	22

Fig. 4.14

Definition 4.9 **Join**

Let

$\mathbf{R} = \{\mathbf{r}^k = \langle r^k_1, r^k_2, .. \, r^k_p, .. \, r^k_n \rangle\}$, $\mathbf{S} = \{\mathbf{s}^i = \langle s^i_1, s^i_2, .. \, s^i_q, .. \, s^i_m \rangle\}$.

Join of relation **R** *on attribute* $\mathbf{r}_p$ *with relation* **S** *on attribute* $\mathbf{s}_q$ *is a relation* **Q** *that contains those concatenated pairs* **r** *and* **s**

$\mathbf{Q} = \mathbf{join}\ (\mathbf{R}, \mathbf{S} : [\mathbf{r}_p \ominus \mathbf{s}_q]) =$

$$\{\mathbf{r}^k \sim \mathbf{s}^i : \mathbf{r} \in \mathbf{R}, \mathbf{s} \in \mathbf{S} \text{ and } (r^k_p \ominus s^i_q) = \text{true}\}$$ *provided* p^{th} *attribute of* **R** *is* **Θ-comparable** *with* q^{th} *attribute of* **S**.

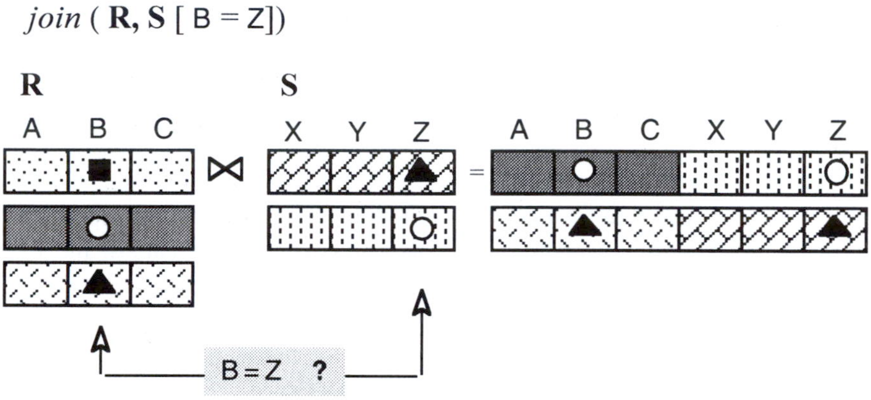

join (**R, S** [B = Z])

R
S
A B C
X Y Z
A B C X Y Z
B = Z ?

Fig. 4.15 Schematic representation of *join*

Conceptually, *join* can be thought of as the *product* of two relations followed by *restriction* subject to a condition specified in the restriction clause; note that the attributes involved in the condition must necessarily be **Θ-comparable**. When joining relations over a number of attributes, the restriction clause will be a conjunction of that number of the relevant simple conditions.

Join is a very powerful operator and together with *project* form a base for normalization - a theoretical support for designing database relations. Most of the join operations are carried out with the restriction clause specified as an equality (as indeed the one presented in Fig. 4.12 - **join** (**R**,**S** : [B=Z]). This kind of *join* is called *equi-join*. The result of *equi-join* is a relation that contains all the attributes from the original two relations. Clearly, at least one attribute is redundant (either B or Z in the above example). The operation of *natural join* (i.e. *equi-join* followed by projection that removes duplicate attributes) may therefore be used, thus saving on somewhat superfluous projections in retrieval programs. From this point on, every join operation in this book is to mean a *natural join*, unless otherwise stated.

Example 4.4

For the database as defined in Example 4.1, express the following query in relational algebra: *Find the location of all non-fiction books giving their authors and titles.*

The following sequence of algebraic operations gives the required answer:

S_1 := *select* SUBJECT (NAME = 'Non-Fiction')

S_2 := *join* (S_1, INDEX: [S_1.CLASS# = INDEX.CLASS#])

S_3 := *project* S_2 (AUTHOR, TITLE, SHELF#)

S_3

AUTHOR	TITLE	SHELF#
King	Strength to Love	24
Hemingway	Death in the Afternoon	22

Fig. 4.16

Note that an equivalent (and more transparent) form for the above query is:

project(
join(
(*select* SUBJECT (NAME = 'Non-Fiction'), INDEX: [CLASS# = CLASS#])
(AUTHOR, TITLE, SHELF#)))

which does not require any auxiliary structures to hold intermediate data.
The remaining operation of relational algebra - *division* - requires the concept of an image set to be defined first.

Definition 4.10 Image Set

Let $\mathbf{R}$ *be a binary relation* $\mathbf{R} = \{\mathbf{r}^k = <x^k, y^k>\}$. *The image set of a particular* x^i *in the relation* $\mathbf{R}$ *is a unary relation containing all those elements* y^p *such that every pair* $<x^i, y^p>$ *belongs to* $\mathbf{R}$, *i.e.* $\mathbf{I}_R(x^i) = \{y^p: <x^i, y^p> \in \mathbf{R}\}$; i, p $\{1,...k\}$

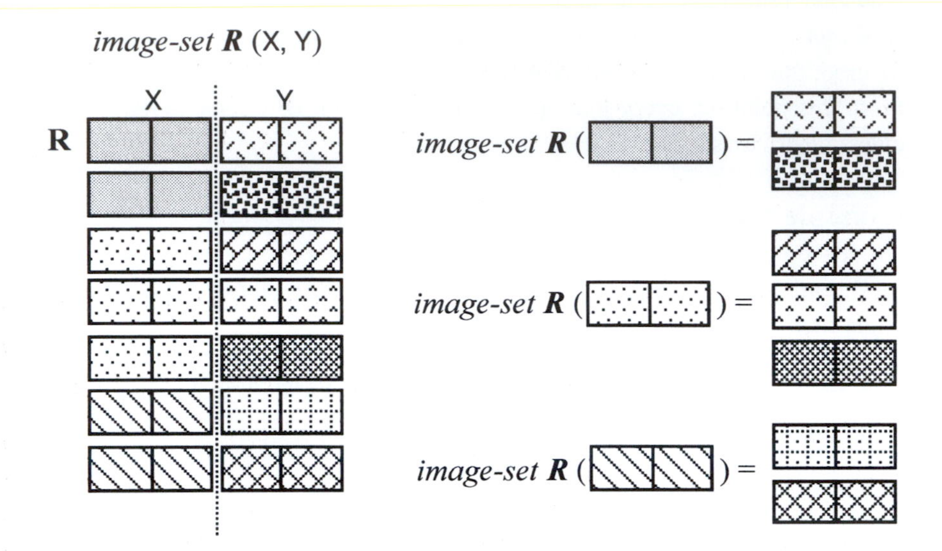

Fig. 4.17 The concept of *image set*

R

MODULE	STUDENT	GRADE
8049	John	B
8049	Mary	A
8049	Sue	B
8050	John	A
8050	Dave	B
8050	Bob	B
8050	Jane	A
8050	Mark	C
8023	Sue	B
8023	John	B
8023	Paul	A
8023	Eve	B

image-set ***R*** (8049)

STUDENT	GRADE
John	B
Mary	A
Sue	B

image-set ***R*** (John)

MODULE	GRADE
8049	B
8050	A
8023	B

image-set ***R*** (8050, A)

STUDENT
John
Jane

Fig. 4.18 Exemplification of image set

Definition 4.11 **Division**

Let $\mathbf{R} = \{\mathbf{r}^i = <x^i_1, x^i_2, ..., x^i_m, y^i_1, y^i_2, ..., y^i_n>\}$.
Let X denote a set of all x*-attributes and* Y *denote a set of all* y*-attributes of* **R**, *thus*
$\mathbf{r}^j(X) = <x^j_1, x^j_2, ..., x^j_m>$ *and* $\mathbf{r}^j(Y) = <y^j_1, y^j_2, ..., y^j_m>$.
Let $\mathbf{S} = \{\mathbf{s}^k = <z^k_1, z^k_2, ..., z^k_n>\}$; *thus* $\mathbf{s}^l(Z) = <z^l_1, z^l_2, ..., z^l_m>$. **R** *is called a* ***dividend*** *relation and may be written in binary form* **R** = {X, Y} *where* Y *contains attributes to be divided; dividing attributes of* **S** *are contained in* Z, *thus* **S** *may be regarded as a unary relation* **S** = {Z}.
Division of **R** *on* Y *by* **S** *on* Z *is defined as:*
divide (R, S: [Y | Z]) = $\{\mathbf{r}^p(X)\text{: } \mathbf{r} \in \mathbf{R} \text{ and } \mathbf{S}(Z) \subseteq \mathbf{I_R}(\mathbf{r}^p(X))\}$
provided that attribute collections Y *and* Z *are union compatible.*

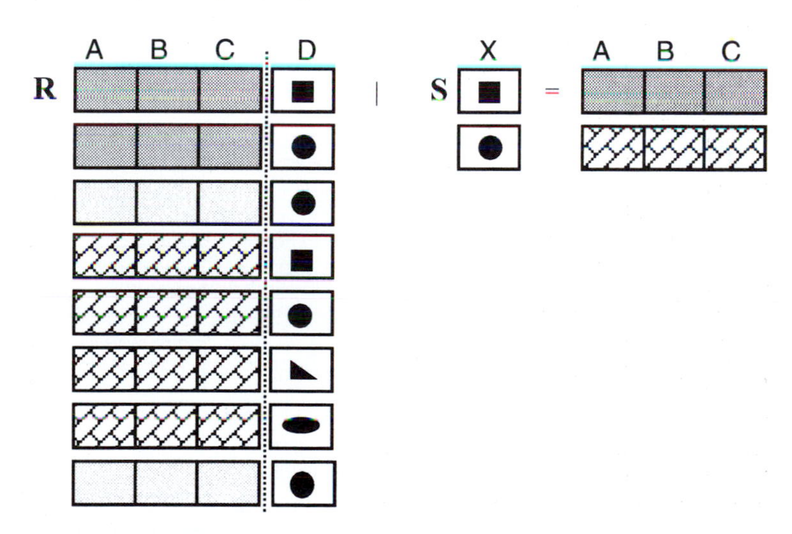

Fig. 4.19 Schematic representation of *division*

Example 4.7

The mechanism of division is demonstrated below by showing all the computations needed to answer a query *Find all the lecturers who teach a course on Prolog* set against the relation **LC** whose possible instance is depicted in Fig. 4.20.

LC

LECTURER	MODULE
Brown	Compilers
Brown	Databases
Green	Prolog
Green	Databases
Lewis	Prolog
Smith	Databases

| **Q**

SUBJECT
Prolog

Fig. 4.20

Following Definition 4.11, the above query can be expressed in terms of division as

P = *divide* (**LC**, **Q:** [MODULE | SUBJECT]) =

=$\{\mathbf{r}^n(\text{LECTURER}): \mathbf{r} \in \mathbf{R} \text{ and } \mathbf{Q}(\text{SUBJECT}) \subseteq \mathbf{I}_{\mathbf{LC}}(\mathbf{r}^n(\text{LECTURER}))\}$

The computations required by the definition are as follows:

r^1(Lecturer) = Brown	►	I_{LC}(Brown) =	Compilers Databases	
r^2(Lecturer) = Green	►	I_{LC}(Green) =	Prolog Databases	√
r^3(Lecturer) = Lewis	►	I_{LC}(Lewis) =	Prolog	√
r^4(Lecturer) = Smith	►	I_{LC}(Smith) =	Databases	

Hence the result relation **P** is composed of two tuples:

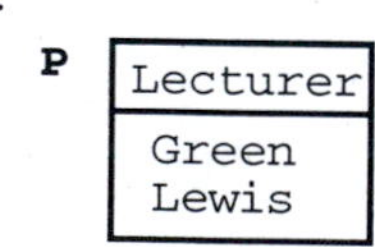

Fig. 4.21

4.3 QUERIES AS COMPOUND ALGEBRAIC EXPRESSIONS

The purpose of this section is to demonstrate how the primitive operations of relational algebra can be applied in practice to represent possible users queries. To be able to do so, we consider a fictitious database that contains information on various goods delivered to a number of branches of a (equally fictitious) Fashion Garments Co (FGC). The conceptual model of the FGC database consists of three relations (primary keys are underlined):

GOODS (Producer, ProductCode, Description)
DELIVERY (Producer, ProductCode, Branch#, Stock#)
STOCK (Branch#, Stock#, Size, Colour, SellPrice, CostPrice, DateIn, DateOut)

Fragments of possible instances of the above relations are depicted in Fig. 4.22.

GOODS

Producer	ProductCode	Description
Moderna Ltd	J612	Jeans
Moderna Ltd	K443	T-shirt
CareFree	J612	Skirt

DELIVERY

Producer	ProductCode	Branch#	Stock#
Moderna Ltd	J612	L1	1003
Moderna Ltd	J612	NY	1003
CareFree	K443	L1	1010

STOCK

Branch#	Stock#	Size	Colour	Sell Price	Cost Price	Date In	Date Out
L1	1003	M	Blue	15.50	15.50	2/5/99	2/5/99
L1	1004	S	Blue	15.50	15.50	2/5/99	8/5/99
NY	1003	L	White	5.50	4.00	3/4/00	Instock
NY	1045	L	Red	17.45	7.85	5/4/00	2/5/00
NY	1020	L	Blue	17.45	7.85	4/5/00	Instock
OX	1004	M	Green	21.30	12.45	6/7/00	8/8/00

Fig. 4.22 FGC database

Query 4.3.1
Find all producers who supply goods
P1 = **project** GOODS(Producer)

Query 4.3.2
Find all producers who have delivered goods to any branch of the company
P2 = **project** DELIVERY(Producer)

Query 4.3.3
Find **Sell-Price** *and* **Cost-Price** *of all goods delivered to branch* **L1** *still in stock*
P3 = **select** STOCK(Branch = 'L1' **and** DateOut = 'InStock')
P4 = **project** P3(SellPrice, CostPrice)

Note that the expression:
project(
select STOCK(Branch = 'L1' **and** DateOut = 'InStock') (SellPrice, CostPrice))

represents the same query without referring to any intermediate structure.

Query 4.3.4
Find **Producer, Product-Code, Description** *for all goods sold at the same day they arrived at any branch*
R1 = **restrict** STOCK(DateIn = DateOut)
R2 = **join** (R1, DELIVERY: [R1.Branch# = DELIVERY.Branch# **and**
R1.Stock# = DELIVERY.Stock#])
R3 = **join** (R2, GOODS: [R2.Producer = GOODS.Producer **and**
R2.ProductCode = GOODS.ProductCode])
R4 = **project** R3 (Producer, ProductCode, Description)

The logic of execution of the above query can be represented by the tree shown in Fig. 4.23. We can thus notice that tree representation of algebraic expressions correspond directly to their functional form; the considered query could be written as:

R4 = **project** (**join** GOODS,
(**join** DELIVERY, (**restrict** STOCK(Date-In = Date-Out):
[Branch# = Branch# **and** Stock# = Stock#]):
[Producer = Producer **and** Product-Code = Product-Code])
(Producer, Product-Code, Description)

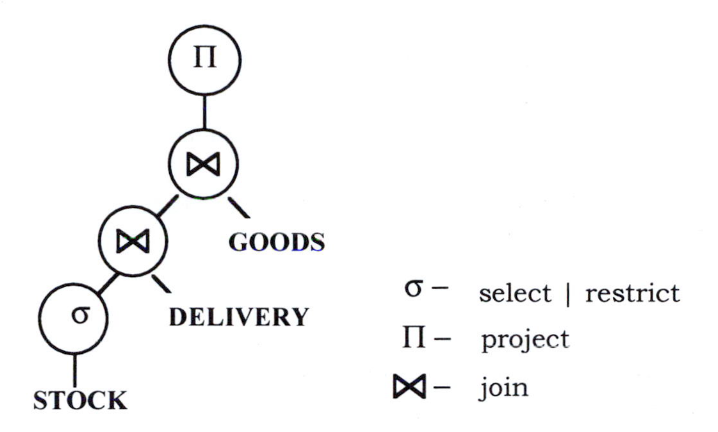

Fig. 4.23

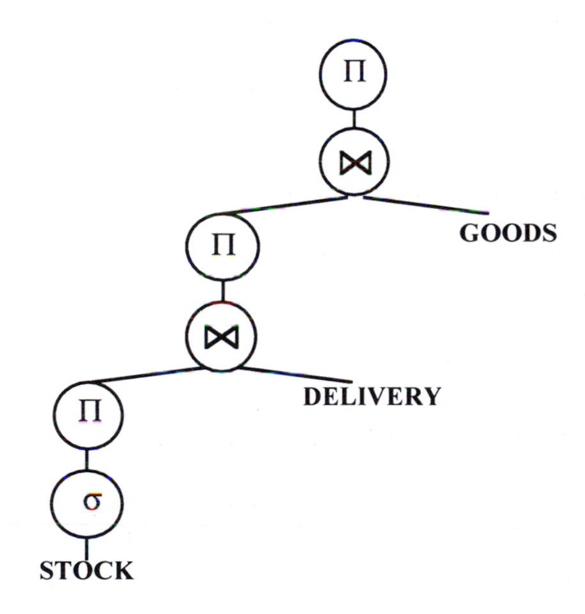

Fig. 4.24

Observing the query tree we can also infer that its execution would not be particularly efficient since attributes that are not required for the final answer (and not needed for further join operations) are nonetheless kept until final projection occurs. Efficiency of algebraic expressions is not necessarily the subject of our considerations at this point, but it is easy to notice that the following sequence (with its tree representation shown in Fig. 4.24) would not have this deficiency:

R1 = **restrict** STOCK(DateIn = DateOut)
R2 = **project** R1(Branch#, Stock#)
R3 = **join** (R2, DELIVERY: [Branch# = Branch# **and** Stock# = Stock#])
R4 = **project** R3 (Producer, Product-Code)
R5 = **join** (R4, GOODS: [Producer = Producer **and** ProductCode = ProductCode])

The tree in Fig. 4.24 (obtained somewhat intuitively) does not quite represent the optimal form of expression 4.3.4 but it does offer some efficiency gains. We will consider that kind of optimization further in Section 4.4.

Query 4.3.5

Find **Branch#, Size, Colour**, **SellPrice** *for all dresses that have not yet been sold*

Q1 = **select** GOODS(Description = '**dress**')
Q2 = **join** (Q1, DELIVERY: [Q1.Producer = DELIVERY.Producer **and**
Q1.Product-Code = DELIVERY.Product-Code])
Q3 = **project** Q2 (Branch#, Stock#)
Q4 = **join** (Q3, STOCK: [Q3.Branch# = STOCK.Branch# **and**
Q3.Stock# = STOCK.Stock#])
Q5 = **select** Q5(Date-Out = 'In-Stock')
Q6 = **project** Q5 (Branch#, Size, Colour, Sell-Price)

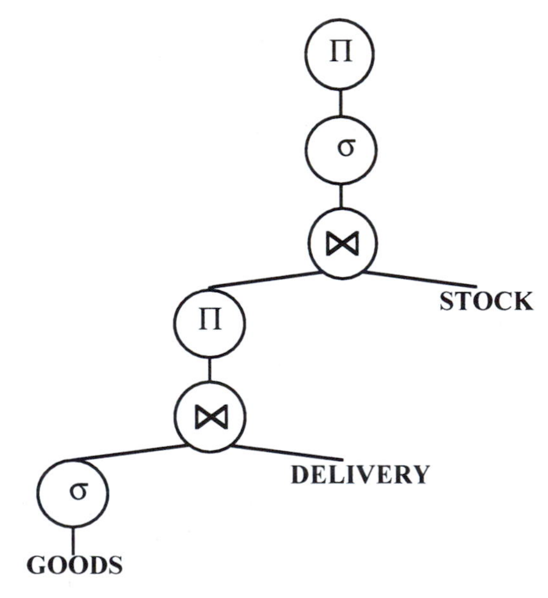

Fig. 4.25

4.4 OPTIMIZATION OF ALGEBRAIC EXPRESSIONS

One and the same computational process can be expressed in a variety of forms that are completely equivalent to each other in the sense that from the identical input they will produce the identical output. This is true for all programming notations and relational algebra is not an exception. However, different forms of code may attract different execution time. Compound algebraic expressions are built from primitive operations (*select*, *project* , *join*, etc.), themselves programmed in some imperative language. Developing fast algorithms for algebraic primitives is far from trivial and we will consider this issue later in this chapter. For now, it will suffice to realize that unary operations require repetitive inspection of every tuple of the operand while binary operations imply some form of combinatorial comparison of tuples from both operands, as Fig. 4.26 illustrates.

selection: **σR**(α) → **R'**	projection: **ΠR**(A) → **R'**

empty (R')
get-tuple (r)
while not eof(R)
    if α **then** R' := R' ∪ r
    get-tuple (r)
end while
	empty (R')
	get-tuple (r)
	while not eof(R)
	  reduce (r, A)
	  if r not_in (R") **then** R' := R' ∪ r
	  get-tuple (r)
	end while

join: **η(R, S)** → **P**

    η(**R**, **S**) ::= **σ(product (R, S))**(join-condition)

```
empty (P)
  while not eof(R)
        x:= r
        while not eof(S)
              x := concatenate (x, s)
              if (join-condition) then P := P ∪ x
              next (s)
        end while (S)
        next (r)
  end while (R)
```

Fig. 4.26

Programmes in relational algebra operate on large sets of data, and thus the order in which the component operations appear in algebraic expressions may have significant impact on the execution time, particularly because the set-oriented paradigm necessarily conceals the details of the actual mechanisms of execution. Relational algebra, being a formal system, offers the usual benefits that come from its mathematical underpinning, that is the existence of equivalence formulae whose veracity can formally be proved and thus they may be freely used to manipulate algebraic statements. Let us consider some of these formulae that are most relevant to optimization. The use of them will shortly be demonstrated on an example query set against the previously described University database.

• **functional composition of selections**

$$\sigma(\sigma R(\alpha))(\beta) \equiv \sigma R(\alpha \cap \beta)$$

σ_α — σ_β — R $\Leftrightarrow$ $\sigma_{\alpha\cap\beta}$ — R

Two nested selections on the conditions α and β respectively can be replaced by a single selection on $\alpha\cap\beta$. The benefit here results from a smaller number of iterations (see Fig. 4.26) since the second loop applied on the result from the first one would no longer be needed; evaluation of the conjunction $\alpha\cap\beta$ takes only insignificantly more time than that necessary to evaluate them separately.

• **functional composition of projections**

$$\Pi(\Pi R(A))(B) \equiv \Pi R(B) \text{ iff } A \supseteq B$$

Π_A — Π_B — R $\Leftrightarrow$ Π_B — R $A \supseteq B$

Two nested projections over a collection of attributes A and B respectively can be replaced by a single projection over the attribute B **wholly contained** within A.

• **commutativity and associativity of join**

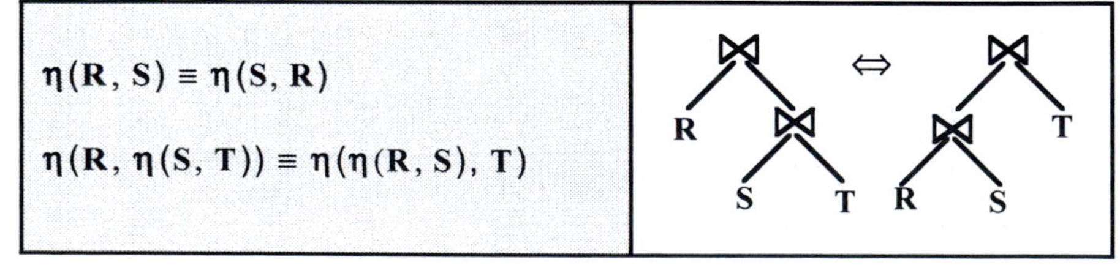

The order in which any two relations are joined is immaterial; consequently any two nested joins can associate in any order. While on their own these formulae do not yield any particular benefit in terms of execution time (save, perhaps, in the cases where the operand relations are located at different sites), they are nevertheless quite useful in manipulating larger expressions.

• **selection before join**

$$\sigma\eta(R, S)(\alpha) \equiv \eta(\sigma R(\alpha), \sigma S(\alpha))$$

Join is a very expensive operation since it involves a nested loop (see Fig. 4.26) on the respective operand relations and hence the cost of execution is proportional to ***cardinality***(R) * ***cardinality***(S). Therefore, rather than performing a selection on the result of join, multiple selections on the operands involved are carried out first (thus reducing the cardinality of both) and then the results are passed to join. As we shall see later the saving on execution time can be quite substantial.

• **projection before join**

$$\Pi\eta(R, S)(A) \equiv \eta(\Pi R(A), \Pi S(A))$$

Projection naturally reduces the size of tuples but it may also diminish the cardinality of join operands (since it removes duplicates). Thus the benefits are of a similar nature to these of the previous case.

• **selection before union/intersection**

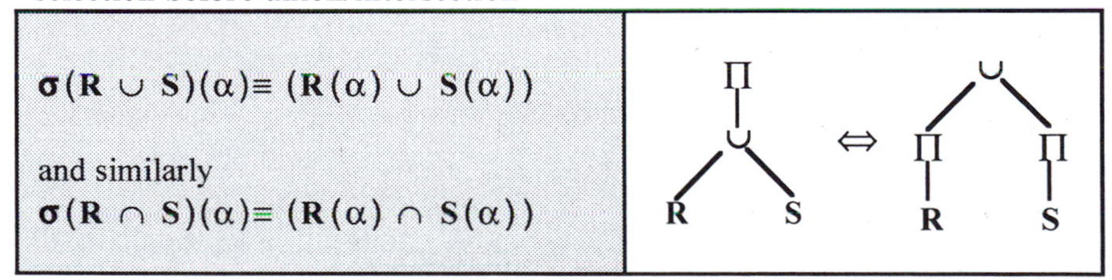

As in the case involving join, selection on the operand relations reduces their cardinality (by removing the tuples not needed for the result anyway) before union (intersection) is called.

Example 4.8

With regard to the following database:

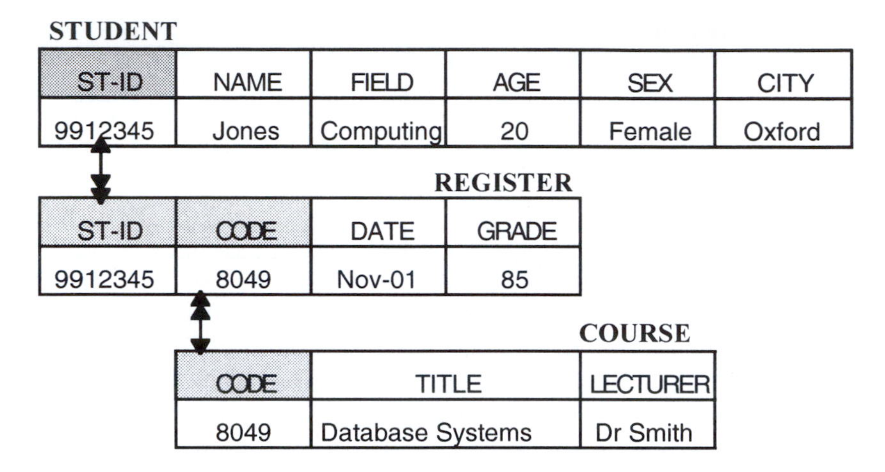

STUDENT

ST-ID	NAME	FIELD	AGE	SEX	CITY
9912345	Jones	Computing	20	Female	Oxford

REGISTER

ST-ID	CODE	DATE	GRADE
9912345	8049	Nov-01	85

COURSE

CODE	TITLE	LECTURER
8049	Database Systems	Dr Smith

the query

Get the names of all female students who live in London and enrolled on the course on database systems

could be written as:

```
project (
   select (
         select (
             select (
                  join (
                      join (REGISTER, COURSE: [CODE = CODE]),
                  STUDENT: [ST-ID = STID])
             (SEX = Female))
         (TITLE = Database Systems))
   (CITY = London))
(NAME)
```

with its corresponding graphical representation:

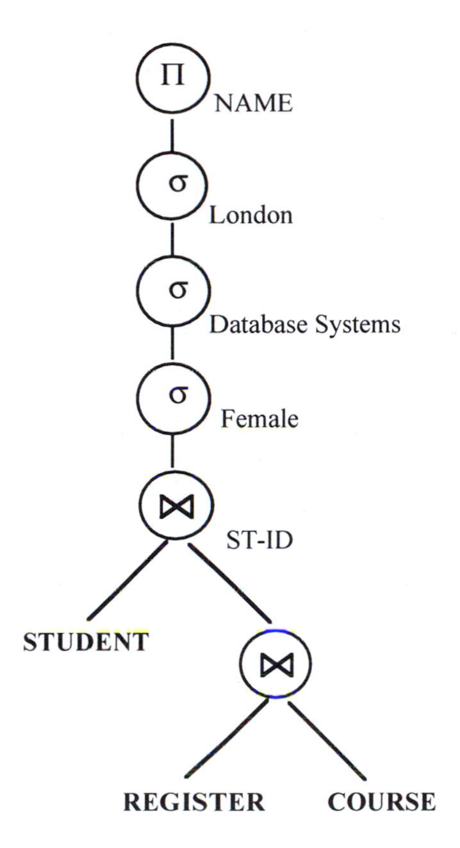

Fig. 4.27 An example of the query tree

Execution of this query in the way implied by the above tree would have been very inefficient indeed. The join operations would necessarily involve a great many tuples: the first one in the region of say 10^5 for REGISTER multiplied by 10^3 for COURSE -(i.e. 100 million attempts to match the tuples referencing all active registrations on modules!), the next one (assuming all courses are taught to some students and the total population is say 20000) would deal with 10^8 times $2 * 10^4$ (i.e. $2*10^{12}$) attempts and then passed a very large table (in practice the entire database held in one table) to the three consecutive selections. Clearly a better execution plan should be available.

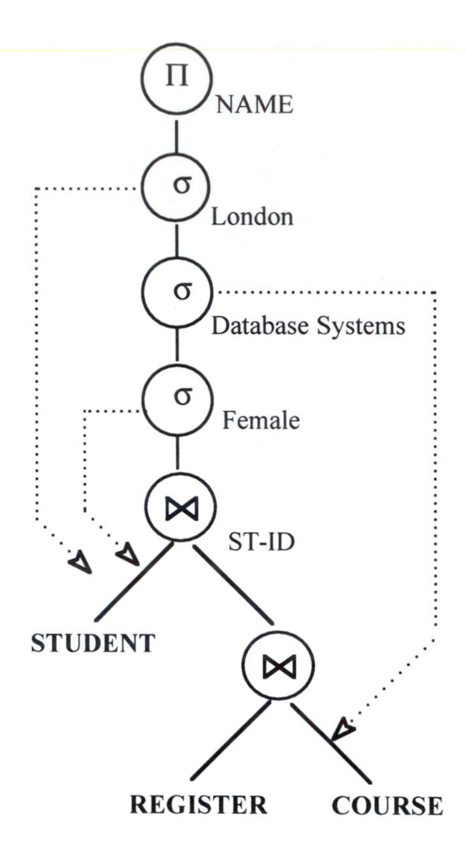

Fig. 4.28 Pushing selections down the query tree

We can observe (see Fig. 4.28) that the three selections with regard to the values of CITY. TITLE and SEX can be pushed down the tree (i.e. carried out before joins) as permitted under the rules described on pages 76 - 77. Modifying the tree in such a way would considerably reduce the number of tuples coming from the respective relations. Out of many tuples in COURSE only one (that describing the Database System course) would now be passed to further processing and, similarly, rather than all tuples from STUDENT only those referring to female students living in London would be involved in the respective join (if half of the students are female and say 20% live in London the number of tuples would be in the region of $2*10^3$). Hence the final join would perform around $2*10^8$ matches which means that the theoretical response time would be 20 thousand times shorter than that implied by the original tree.

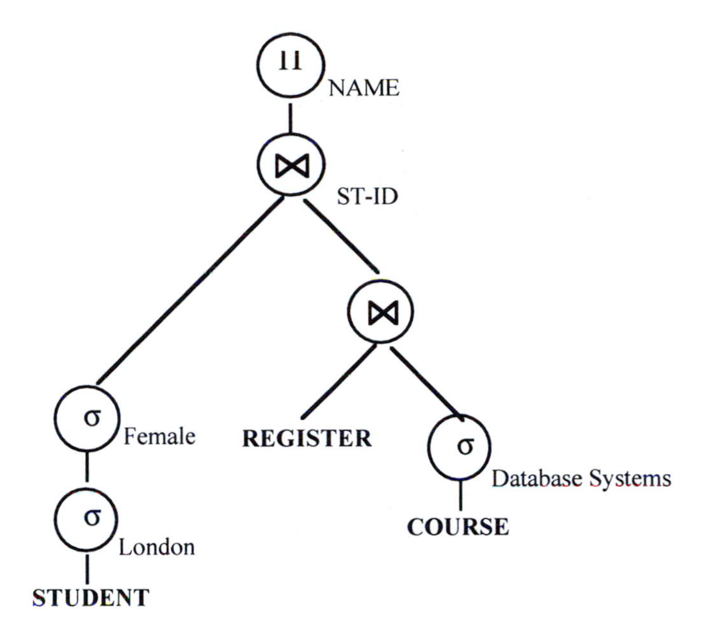

Fig. 4.29 Modified query tree

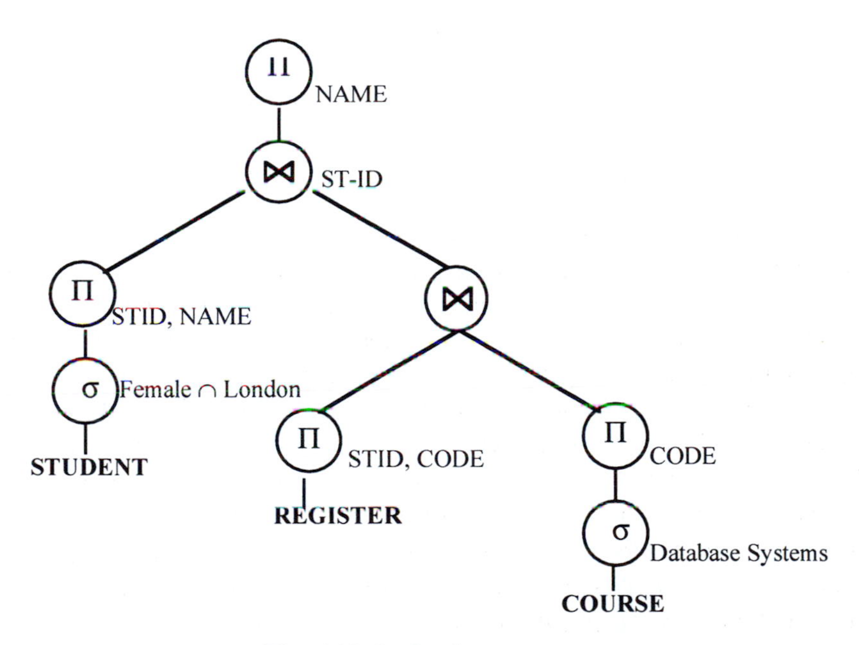

Fig. 4.30 Optimal query tree

Further improvements can be achieved by handling projections. Clearly, several attributes can be discarded at the early stage of processing since they are needed neither for the final answer (only student names are required) nor for any intermediate operations (as joins are made only with regard to STID and CODE). The final step (see Fig. 4.30) would be to replace two consecutive selections by a single one subject to conjunction of their individual conditions.

Optimization as illustrated on the above example is a semi-heuristic process since at every stage of transformation: *previous-tree* $\rightarrow$ *rule* $\rightarrow$ *better-tree* there is a certain amount of indeterminacy as to the choice of rules. In consequence, for a given query many logically equivalent processing plans can be derived.

In practice, optimization is carried out automatically by a program (query planner) integrated with the DBMS whose existence (let alone details) is concealed from application programmers. In any case, database applications are typically developed in a very high-level language (such as SQL, considered later in this book) where no means of specifying any navigation (whether in terms of data structures or processes) are made available to the programmer.

Another aspect of efficient algebraic computation are the algorithms through which the primitive operations are implemented. The algorithm for join is of particular importance since that operation is very computationally expensive. The simplest (and the slowest) is presented in Fig. 4.26. Essentially it attempts to match every tuple from one operand with every tuple from the other; hence the cost.

Nearly all improved algorithms, however, attempt some form of (logical) partitioning of the operands into fragments that contain only the tuples susceptible to joining (i.e. they conform to the join condition). This pre-processing of operands is typically achieved through techniques such as sorting, indexing or hashing; compatible fragments are then joined via the nested loop method.

An example of this approach is the *sort-merge* algorithm whose mechanism is presented using data structures shown in Fig. 4.31. The operands R and S are held in linked lists sorted on the fields over which join is made, that is A and B respectively; the results are stored in a similar list T. For clarity, the lists are shown as if they were physically sorted though the same effect (or indeed better) can be achieved via logical indexing. This ordering of the relations R and S has on its own partitioned them into groups of tuples holding identical values for the attributes A and B.

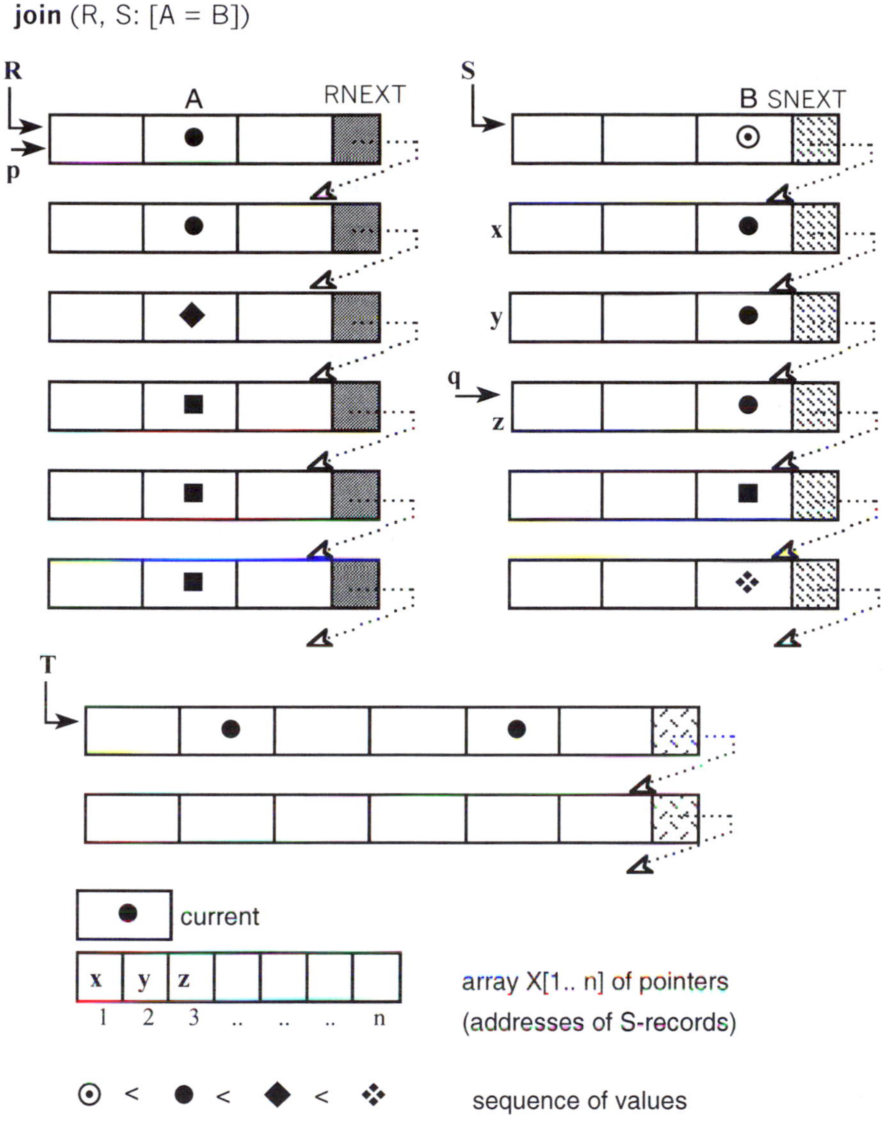

join (R, S: [A = B])

R

A RNEXT

p

S

B SNEXT

x

y

q

z

T

current

x	y	z				
1	2	3	..	..	..	n

array X[1.. n] of pointers (addresses of S-records)

⊙ < ● < ◆ < ❖ sequence of values

Fig. 4.31 Data structures for *sort-merge* algorithm for join

4.3 Consider the following three relations:

TAPE (<u>Catalogue #</u> ,Title, Purchase-Price, Rent-per-Day, Date-Bought)
MEMBER (<u>Id,</u> Name, Address, Join-Date)
BORROWING (<u>Member-Id, Catalogue#, Borrow-Date,</u> Return-Date)

Use Relational Algebra to represent the following queries:
(a) What different titles does the club have?
(b) Does the club have a video tape entitled Quadrophenia?
(c) Is a copy of Quadrophenia available for rental?
(d) Who has borrowed Paradise Lost?
(e) List all the titles that have not been borrowed for the last 2 years.

4.4 Prove that **select (restrict R**(b))(a) = **restrict (select R**(a))(b) for all valid a and b

4.5 Prove that for any relation **R**(X,Y)

select (select R($X=x_0$) ($Y=y_0$)) = **select** (**R**($X=x_0$ **and** $Y=y_0$))

What is an equivalent form of **select** (**R**($X=x_0$ **or** $Y=y_0$)) ?

4.6 Prove that division is not strictly necessary, i.e. can be replaced by a combination of some other relational algebra operations.

4.7 Derive formally (i.e. by showing all manipulations and justifying every transformation) the query processing plan described in Example 4.8

4.8 In the *sort-merge* algorithm as presented above the **q**-pointers are stored in a one-dimensional array. Devise a better structure for holding these addresses, modify the algorithm and pseudocode as appropriate, and explain the benefits arising from your suggestion.

CHAPTER 5

LEAP - the algebraic DBMS

5.1 INTRODUCTION

Chapter 4 presented relational algebra as a theoretical notation for manipulating relational structures in a way that naturally conforms to the principles of the relational model. However, relational algebra has never been a natural choice for a language that could be part of DBMS, for algebraic expressions were considered too distant from the conventional programming style. Thus, the principles of relational algebra were used to develop a more habitual form of a programming language, SQL, which then became the *de facto* standard for database applications; its quite detailed description is given in Chapter 8.

Yet, relational algebra may be used directly as a sole means of communication with the DBMS! Led by this somewhat obvious conclusion one of the authors of this book designed and implemented a simple educational DBMS - Leyton's Extensible Algebraic Processor, or LEAP, for short. The system was designed with clear didactic objectives; it meant to provide a learning tool whereby the students could gain practical experience in formulating algebraic expressions, observing how they run and, by inspecting and analysing the source code, becoming familiar with various aspects of developing relational DBMSs.

LEAP is an open source project, distributed free under the terms of the GNU General Public Licence. It has been developed in C on a GNU/Linux Redhat system, and is known to run on Linux, Solaris, SunOS, HP-UX, Ultrix, IRIX, AIX, OSF1, BeOS, Free BSD, and Windows 95/NT. The book's accompanying website provides the relevant details for downloading, installing and running the system. LEAP comes with the full source code, extensively documented and commented and the relevant manuals. LEAP has been successfully used to support various courses on databases run at some 100 different sites worldwide.

This chapter discusses some of the basic design and implementation issues relating to LEAP, and will draw on examples from the previous chapter to demonstrate how it can be used as an aid to learning and understanding the relational theory.

5.2 LEAP ARCHITECTURE

A database management system is fully relational when it supports three fundamental aspects of the model: the structures, general integrity constraints and operations (a data sublanguage at least as powerful as relational algebra). Due to its assumed scope (essentially to make the system simple), LEAP differs, on occasions, from some of the established relational principles. It is worth bearing in mind, however, that no system (to the authors' knowledge) can claim to be fully relational in every aspect. Compromises are invariably necessary when translating a theoretical design into a working model.

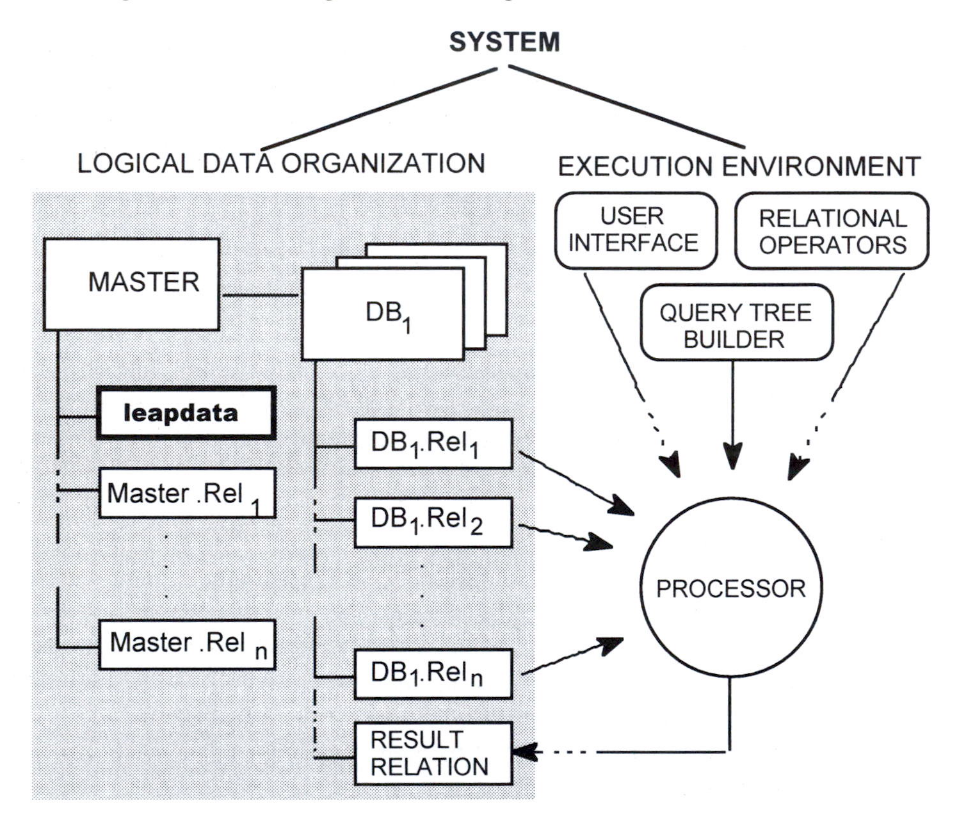

Fig. 5.1 LEAP architecture

The system's logical architecture is exemplified in Fig. 5.1. At the highest level of abstraction, LEAP is composed of two parts - the logical data organization and the execution environment.

5.2.1 Logical data organization

LEAP may contain many databases, that is groups of related relations. For example, we could conceivably have two databases, one containing a bank's employee information, the other containing information about stock market securities that the bank trades.

In addition to the various databases containing user information (which we can collectively term 'user' databases, such as DB_1, DB_2, etc.), there is a separate database containing information about the other databases (forming a simple data dictionary). This is the MASTER database. In all respects its structure is the same as that of other databases, except for the information content held (naturally) in relations called system relations.

When LEAP starts, the MASTER database is opened first. LEAP is programmed with the location references of the various files that represent its relations. The **leapdata** relation holds the location references of the 'user' databases. Each database is held in a separate directory and can be opened (one at the time) upon request from the user.

Every relation in LEAP comprises two files: the **hash file** used to enforce the uniqueness rule of the tuples in the relation and the **relation file**.

Each relation is composed of a 'header' and a 'body'. The 'header' (attribute names) defines the meaning of the data that is stored in the 'body' (the tuples). Formally, the header is a fixed set of attribute/domain pairs: $\{(A_i:D_i)\}^*$. However, LEAP (along with many database systems) supports domains only through the data types. Thus, the underlying relation file stores for each component attribute its name, data type and size.

The body of a relation consists of a set of tuples, which varies over time. Each tuple consists of a set of attribute-value pairs$\{(A_i:V_{ij})\}^*$, the values being drawn from the appropriate domains. The files that hold the tuples are organized sequentially. Each piece of the data is stored in the same order the attributes were built up at the start of the file and the first tuple starts immediately after the last element of the header.

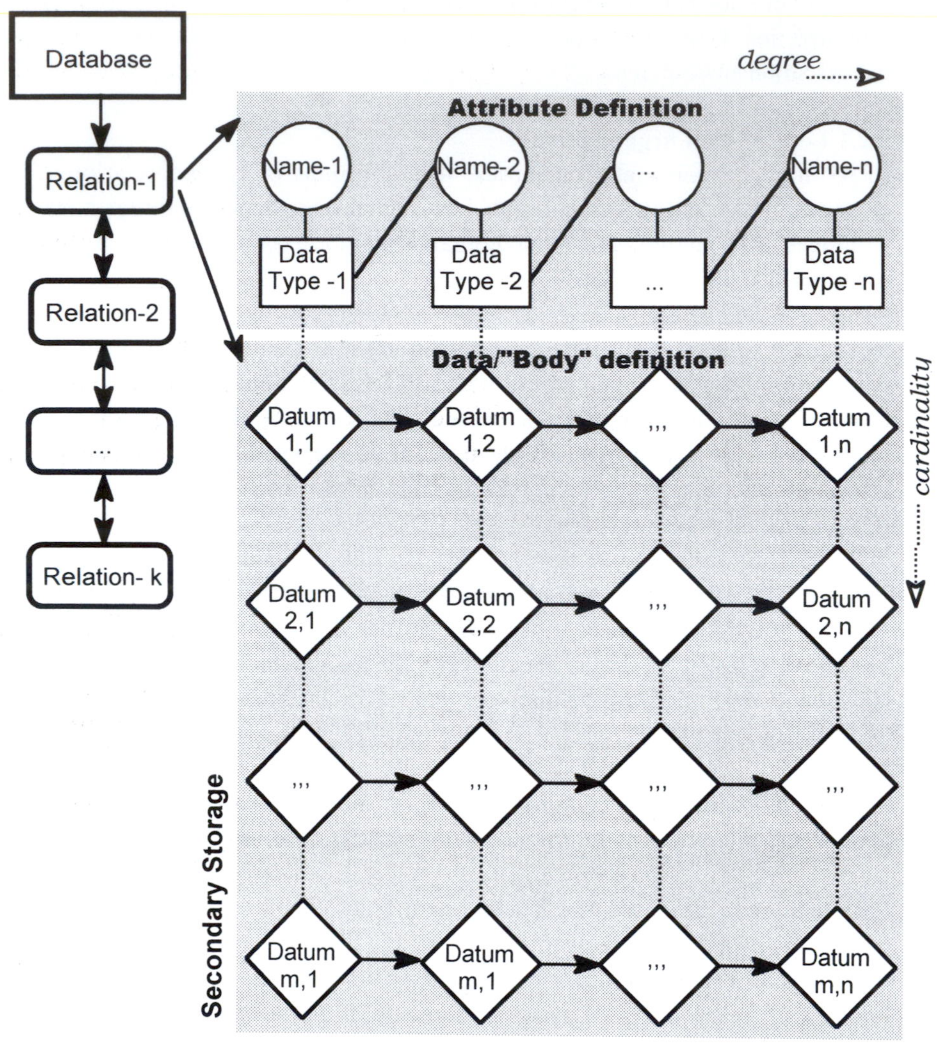

Fig. 5.2 Representation of relations in LEAP

The system does not read in all of the data from the relations in each opened database. This would have been wasteful of time, memory and resources. Once a particular 'user' database is opened, LEAP consults another special data dictionary relation, **leaprel**, held in every database, and lists all of the other relations in the database.

When a tuple is referenced during the execution of relational operators, a structure is set up in the memory. The underlying file is read to populate the information with 'header' information first and, as the operator progresses, subsequent valid tuples are copied from the 'body' file records into the structure. This process works in reverse for writing out a record.

When new data (through an insert or a delete into the relation) occurs, LEAP computes a hash key from the value provided in the tuple, and uses this to check for the uniqueness of that value. This is far more efficient than re-scanning the relation. Only when the tuple is thereby proven to be unique, is it stored in the relation.

As mentioned above, the data dictionary primarily resides in the MASTER database. However, there are other useful pieces of information about the user databases which, though part of the data dictionary, are actually stored within the 'user' databases.

These are:

leaprel	The relations that make up the current user database
leapscripts	The scripts that may be executed
leapattributes	Attribute names, types, and sizes in each relation
leaptypes	The data types available
relship	The relationships that exist between relations

Fig. 5.3 Data dictionary components stored in ‘user’ databases

The relations that form the data dictionary (whether in the MASTER or user databases) are structurally identical to any relation created by the user.

5.2.2 Execution environment

The execution environment is the overall term for the running instance of LEAP. It references and modifies the data through processing user commands, and executing specific operations, as appropriate. The typical result of processing is an additional set of relations or a modification to an existing set of relations.

The execution environment receives from the user interface a command to be processed. This command is not immediately executable since it needs to be checked for lexical, syntactic and semantic correctness, and then converted into an internal representation.

The process, generally referred to as parsing, creates an internal data structure (the **query tree**). Parsing starts from the lexical analysis (verification of spelling and punctuation), followed by syntax analysis (structural correctness of expressions, algebraic or otherwise) and verification of relation and attribute names.

The language of relational algebra as used in LEAP is defined below (Fig. 5.4) in BNF. This section only specifies that part of the language which relates to performing algebraic operations on the relations; for clarity, other (generally non-algebraic) operators are not included here.

```
<command> ::= [<relation name> = ] <relational-exp>
<relation-exp> ::= <unary-term>|<binary-term>|relation-name
<unary-term> ::= project (<relational-exp>) (<attribute-list>) |
                 select (<relational-exp>) (<qualification>)
<attribute-list> ::= <attribute-name> [,<attribute-name>]*
<qualification> ::= <attribute-comp> [{and|or} (<attribute-comp>)]*
<binary-term> ::= (<relational-exp>) <operator> (<relational-exp>) |
              join (<relational-exp>) (<relational-exp>) (<qualification>)
<operator> ::= union | difference | intersect | product
<attribute-comp>::= <attribute-name> <comparator> {attrib-name|value}
<comparator> ::= { < | > | <= | >= | = | <> }
```

Fig. 5.4 BNF definition of relational algebra

The relational operators form the core of the system. The operators are passed all appropriate references to relations, conditions and expressions, and process these to produce a relation. In themselves, they are not aware of the overall context in which they operate. The system ensures that they are called in the correct sequence, with the correct parameters, based on the query tree that was built by the parser. The results from each operators execution must also be properly handled to ensure that nested expressions are executed properly.

The following operators are implemented in LEAP: **project**, **select** (which combines select and restrict), **union**, **difference**, **intersect**, **product**, and **join**. These form a relationally complete set of operators; others (such as image set, division) need to be expressed in terms of the implemented ones. The high-level algorithms for each of these operators is given below in a C-like notation.

Project

```
perform_projection( R, Attribs) {
     Result=new_relation(Attribs);
     Tuple=readfirst_tuple(R);
     While (tuple is valid) {
          New_tuple=fetch_values(tuple,attribs)
          ok=check_hash(new_tuple); /* Check that the hash
                                   does not already contain the new tuple */
          If (ok==FALSE) {
               Write_tuple(result,new_tuple);
               Add_to_hash(new_tuple);
          }
          Tuple=readnext_tuple(R);
     }
     return(result);
}
```

Fig. 5.5

The operator *fetch_values* extracts the appropriate attributes. By referencing a hash structure, writes only occur with new unique values for the resulting relation.

Select

```
Perform_select ( R, Condition ) {
     Result=new_relation(R);
     Tuple=readfirst_tuple(R);
     While (tuple is valid) {
          If (check_condition(Condition,tuple)==TRUE) {
          Write_tuple(result,tuple);
          }
          tuple=readnext_tuple(tuple);
     }
     return(result);
}
```

Fig. 5.6

The process of checking the condition examines the context in which the condition is presented. There are two distinct types of condition evaluation:

- comparison of a named attribute value with a provided value,
- comparison of a named attribute value against another named attribute.

It has been assumed that since the operand is a proper relation, checking for duplicates in the result is not necessary.

Union

```
Perform_union(R, S) {
    If union_compatible(R,S) {
      Result=new_relation(R);
      Tuple=readfirst_tuple(R);
      While tuple is valid {
            Write_tuple(result,tuple);
            Add_to_hash(tuple);
            Tuple=readnext_tuple(R);
      }
      tuple=readfirst_tuple(S);
      while tuple is valid {
              if (check_hash(tuple)==FALSE) {
                    write_tuple(result,tuple);
                    add_to_hash(tuple);
              }
              tuple=readnext_tuple(S);
      }
       return(result);
    } else {
        return(ERROR);
    }
}
```

Fig. 5.7

A hash table for the values in relation R is consulted by the procedure as it processes relation S. For all values that do not exist in the hash table, the result is written to the result relation.

Difference

```
Perform_difference(R,S){
    If union_compatible(R,S) {
        Result=new_relation(R);
        Tuple=readfirst_tuple(S);

          /* Build up a hash of values in S,
          but don't   write them to the result */
          While tuple is valid {
               Add_to_hash(tuple);
               Tuple=readnext_tuple(S);
          }
          tuple=readfirst_tuple(R);
          while tuple is valid {
               /* Does the value exist in hash built from S? */
               if (check_hash(tuple)==FALSE) {
                   write_tuple(result,tuple);
                   add_to_hash(tuple);
               }
               tuple=readnext_tuple(R);
           }
          return(result);
    } else {
          return(ERROR);
    }
}
```

Fig. 5.8

Difference makes similar use of the hash structure for checking whether or not a record exists in the other relation.

Intersection

The logic underlying the operation *intersect* is fundamentally the same as *difference*. The difference is simply in the checking that is done against the hash table. Intersection returns all tuples that exist in *both* relations R and S and thus the comparison would read:

```
if (check_hash(tuple)==TRUE)
```

rather than:

```
if (check_hash(tuple)==FALSE)
```

Product

```
Perform_product(R,S) {
    Result=new_relation(R,S);
    RTuple=readfirst_tuple(R);
    While Rtuple is valid {
        Stuple=readfirst_tuple(S);
        While Stuple is valid {
            Ctuple=Combine_tuple(Rtuple,Stuple);
            Write_tuple(result,ctuple);
            Stuple=readnext_tuple(S);
        }
        RTuple=readnext_tuple(R);
    }
    return(result);
}
```

Fig. 5.9

The Cartesian product operation is simple but computationally expensive. The resulting relation can quickly become quite large, and there is no significant optimization possible other than perhaps cashing smaller relations in memory.

Join

```
Perform_join_nested( R, S, Condition ) {
    Result=new_relation(R,S);
    RTuple=readfirst_tuple(R);
    While Rtuple is valid {
        Stuple=readfirst_tuple(S);
        While Stuple is valid {
            If (check_condition(Condition,rtuple,stuple)==TRUE){
                Ctuple=Combine_tuple(Rtuple,Stuple);
                Write_tuple(result,ctuple);
            }
            STuple=readnext_tuple(R);
        }
        RTuple=readnext_tuple(R);
    }
    return(result);
}
```

Fig. 5.10

Since LEAP does not support indices, the basic nested loop algorithm was implemented for join. Unfortunately, this requires a large number of scans of the second relation, which slows down the process. However, LEAP is primarily a teaching/learning tool and thus does not usually handle relations large enough to affect adversely the response time. LEAP is a command line, interpreted system. Once the user interface has received a command from the user, a call to the query tree processor ensures that the query is valid. The resulting query tree structure must then be processed appropriately.

The system will move through the query tree in a structured fashion, calling the appropriate operators and ensuring that the resulting relations are passed on as necessary. Where nesting of operators occurs, the result from a particular operator that is to be used by a higher operator is passed up the query tree structure. Thus, the query tree is the link by which all the operators are tied together during execution. In addition to supporting the relational operations, there is also a user environment to support. Such items include:

- creation, deletion and renaming relations
- scripting large sets of commands
- inserting and deleting data from relations
- display and/or printing relations, performance statistics and debug information.

LEAP is an educational rather than industrial DBMS, thus its role is to demonstrate the principles of relational databases. Large industrial systems offer many other features now deemed necessary for data-sensitive or performance-critical applications. These features include concurrency control, recovery and backup, data and process distribution, data security, transaction integrity, SQL support and others.

For illustrative purposes a sample run of LEAP has been enclosed in the next section. The example chosen for that demonstration is essentially identical to Example 4.1 from the previous Chapter. The run was carried out on a Linux system, screen-captured into a file and reprinted with only insignificant editing.

5.3 A SAMPLE RUN OF LEAP

```
[username@somemachine src]$ ./leap
LEAP 1.2.5.1 - An extensible and free RDBMS
Copyright (C) 1997-1999 Richard Leyton.
.........................................
```

```
[user] :-) create example
Resetting all updated flags: Done.
Updating hash tables: ....................
Message: Opening the [example] database....
Resetting all updated flags: Done.
Updating hash tables: ....................

[example] :-) use
Valid databases are:
--------------------
master
user
tempdb
.......
example

[user] :-) use example
Resetting all updated flags: Done.
Updating hash tables: ....................
Message: Opening the [example] database...

[example] :-) list
NAME                      TEMP  SYSTEM
------------------------- ----- ------
leaprel                   FALSE TRUE
leapscripts               FALSE TRUE
leapattributes            FALSE TRUE
leaptypes                 FALSE TRUE
relship                   FALSE TRUE
Message: Relation zzsebs returned.

[example] :-) relation (book)
((reference,STRING,9),(author,STRING,9),(title,STRING,30))
[example] :-) relation (subject)
((class,STRING,5),(class_name,STRING,15))
[example] :-) relation (index)
((author,STRING,10),(title,STRING,30),(class,STRING,5),(shelf,INTE
GER,4))
[example] :-) add (index) (JOYCE,ULYSSES,C1,12)
[example] :-) add (index) (GREENE,SHORT STORIES,C1,14)
....
[example] :-) add (book) (R003,JOYCE,ULYSSES)
[example] :-) add (book) (R004,JOYCE,ULYSSES)
....
[example] :-) add (subject) (C1,FICTION)
[example] :-) add (subject) (C2,SCIENCE-FICTION)
[example] :-) list
```

```
NAME                      TEMP  SYSTEM
------------------------- ----- ------
leaprel                   FALSE TRUE
leapscripts               FALSE TRUE
leaptypes                 FALSE TRUE
relship                   FALSE TRUE
leapattributes            FALSE TRUE
index                     TRUE  FALSE
book                      TRUE  FALSE
subject                   TRUE  FALSE
Message: Relation zzyfyi returned.

[exaxmple] :-) print subject
class class_name
----- ---------------
C1    FICTION
C2    SCIENCE-FICTION
C3    NON-FICTION
C4    SCIENTIFIC
C5    POETRY
C6    DRAMA
Message: Relation subject returned.

[example] :-) print index
author     title                          class shelf
---------- ------------------------------ ----- -----
JOYCE      ULYSSES                        C1    12
GREENE     SHORT STORIES                  C1    14
ORWELL     ANIMAL FARM                    C1    12
LEM        ROBOTS TALES                   C2    23
LEM        RETURN FROM THE STARS          C2    23
GOLDING    LORD OF THE FLIES              C1    12
KING       STRENGTH TO LOVE               C3    24
HEMINGWAY  DEATH IN THE AFTERNOON         C3    22
HEMINGWAY  TO HAVE AND HAVE NOT           C1    12
Message: Relation index returned.

[example] :-) print book
reference author    title
--------- --------- ------------------------------
R003      JOYCE     ULYSSES
R004      JOYCE     ULYSSES
R023      GREENE    SHORT STORIES
R025      ORWELL    ANIMAL FARM
R033      LEM       ROBOTS TALES
R034      LEM       RETURN FROM THE STARS
R036      GOLDING   LORD OF THE FLIES
```

```
R028       KING       STRENGTH TO LOVE
R143       HEMINGWAY DEATH IN THE AFTERNOON
R149       HEMINGWAY TO HAVE AND HAVE NOT
Message: Relation book returned.

[example] :-) a=project (book) (author)
Message: Relation a returned.

[example] :-) print a
author
---------
JOYCE
GREENE
ORWELL
LEM
GOLDING
KING
HEMINGWAY
Message: Relation a returned.

[example] :-) b=(project (subject) (class)) difference (project
(index) (class))
Message: Relation b returned.

[example] :-) print b
class
-----
C4
C5
C6
Message: Relation b returned.

[s] :-) quit
Message: Closing [s] database.
.....
Message: LEAP Terminated successfully!
[username@somemachine src]$
```

5.4 EXERCISES

5.1 Download LEAP from the book's website and install the system on your PC.

5.2. Use LEAP to carry out Exercises 4.1 - 4.3.

CHAPTER 6

Normalization

6.1 DESIGNING RELATIONS

The principle of central control of corporate information laid down the foundations for database theory. The characteristic database features (data shared amongst various applications, controlled redundancy, data independence from processing) are all derived from this principle. Inevitably, the development of relational database theory has centred around the issues of store and processing economy. An optimal structure of sets of data has been sought.

Before we proceed any further we reiterate again what we understand by a relational database. It is a system that is perceived as a set of time-varying n-ary relations (and nothing but relations) whose behaviour is guarded by the general integrity rules. All accesses to the database, whether active or passive, are expressed in terms of set-level operations (Relational Algebra or equivalent) in a way that any transformation of relations preserves the principle of relational closure.

From a theoretical standpoint two questions emerge quite naturally:

- *Given a set of relations, say* $\{R_i\}$, *can we represent it by another set of differently structured relations* $\{S_j\}$?
- *Which set is better, and subject to what criteria* ?

From a more practical point of view, by observing behaviour of some data we can see that the relations holding them may attract certain undesirable properties - the so-called insertion, deletion and update anomalies. We shall consider these anomalies in considerable detail, but at the moment it suffices to say that, whenever any of the anomalies occur, updating the corresponding relations may prove difficult or indeed impossible.

The problem then is to find better relations that are free from these anomalies, thus ensuring consistency of the data in the database. The optimal structure must, of course, be capable of carrying exactly the same information content as the original one and any possible increase of data volume due to structure transformation is to be minimized. We are then faced with a typical design problem.

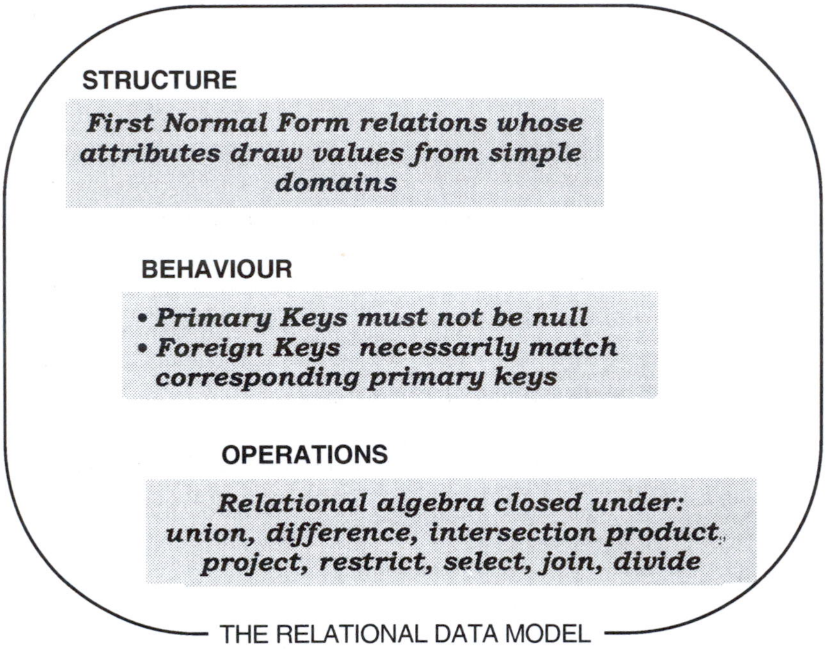

STRUCTURE

First Normal Form relations whose attributes draw values from simple domains

BEHAVIOUR

- *Primary Keys must not be null*
- *Foreign Keys necessarily match corresponding primary keys*

OPERATIONS

Relational algebra closed under: union, difference, intersection product, project, restrict, select, join, divide

THE RELATIONAL DATA MODEL

Fig. 6.1 The components of the relational data model

A general method to solve that problem is called **normalization**. In relational theory normalization can be regarded as a discrete finite algorithm that produces a family of relations - all derived from some initial set of relations which were obtained by applying (say) EAR modelling techniques. The mechanism of transformation from one structure to the better one is based on particular kinds of internal relationships between attributes; it thus ensures that the final relations are free from the anomalies mentioned earlier.

It must be stressed here that this algorithm, though always halting successfully in a finite number of steps, can actually produce different results for the same set of original relations. The overall idea is to decompose the initial relations into conceptually simpler ones for as long as the resulting structures express one simple concept each. However, since this decomposition can be done (properly) in a variety of ways, the results may differ.

Normalization theory assumes that an appropriate Data Sub-Language (Relational Algebra or equivalent) guarantees executability of any process, be it retrieval, updating or restructuring. However, with respect to the data being independent of processing, none of these processes is necessarily considered in detail when the data model is being designed.

Secondly, each data model is considered by its developer to be structurally stable, i.e. independent to some extent over time. Therefore the period in between subsequent restructurings (however complex thy might be) is considered long enough for operational purposes. During these periods the database structure is then in a time-independent state. It does not, of course, mean that users cannot delete some data no longer needed, or insert new values whenever required. By definition all the relations are *time-varying* hence prone to the usual updating operations. What it does mean is that any new relation to be added to the database must be designed according to the normalization principles.

The way normalization is presented in this book aims at highlighting its virtues as a design tool. Therefore all aspects that do not contribute to this process will be treated with less attention.

6.2 FUNCTIONAL DEPENDENCY. BCNF NORMALIZATION

Attributes within a relation depend somehow on each other. These intra-relationships are the result of the semantics of the data, that is they express certain facts about possible connections between various items of data. We have already met a particular kind of such a relationship, namely the concept of key as a unique identifier discussed in Chapter 3, and will now gradually generalize this concept. We first introduce the concept of functional dependency quite informally by considering simple examples, returning to more rigorous definitions later.

Example 6.1

Let DINNER-PARTY describe a party planning exercise as shown in Fig.6.2. We have assumed unique names for the guests and that the food consists of four components: two vegetables, meat and wine.

DINNER-PARTY

GUEST	VEG1	VEG2	MEAT	WINE
Mary	Potato	Tomato	Beef	Red
John	Pasta	Salad	Beef	Red
Eve	Rice	Tomato	Pork	White
Jane	Potato	Salad	Duck	Red
Peter	Pasta	Tomato	Fish	White
Dave	Rice	Salad	Pork	White

Fig. 6.2

By looking at the values of MEAT and WINE we can conclude that whatever the other components of the dinner are, beef is associated with the red wine, pork with the white wine, etc. We can thus say that **the choice of** MEAT **determines** WINE (but not the other way round).

Example 6.2

Let CAR-RENTAL be a relation that holds some details about cars for hire in some company. Fig. 6.3 shows a possible instance of this relation.

CAR-RENTAL

MAKE	MODEL	ENGINE	COLOUR	CATEGORY
Volvo	740	2.0	Blue	A
Honda	Civic	1.6	Red	C
Rover	Mini	1.1	Green	D
BMW	320	2.0	Green	A
Ford	Focus	1.8	Black	B
Ford	Kia	1.1	Red	D
Rover	45	1.6	Red	C

Fig. 6.3

To identify a particular car we need three attributes MAKE, MODEL and ENGINE; considering possible further additions to or deletions from the stock of cars no other collection of attributes could serve as a key in this relation.

Again, observing the values of ENGINE and CATEGORY we notice that whenever ENGINE equals 2.0, the corresponding value of CATEGORY is invariably equal to A. Furthermore, the same applies for all values in ENGINE, that is **every value of** ENGINE **has exactly one value of** CATEGORY **associated with it**. The relation holds extra information concerning a company pricing policy, which could have been stated as **'rent charges for any car are determined solely by its engine capacity'.**

Definition 6.1 **Functional Dependency**

> *Let* $\mathbf{R} = \{\mathbf{r}^i = <r^i_1, r^i_2, ..., r^i_n>$ *where* $i \in I = \{1, 2, ...\}\}$. *Let* X *denote a set of all the attributes of* **R**, *thus*
> $\mathbf{r}^j(X) = <r^j_1, r^j_2, ..., r^j_n>$, *and* Y, Z *be some subsets of* X.
> *We say that functional dependency* $Y \rightarrow Z$ *holds in* **R** *if whenever two tuples, say* j *and* k, *agree in value of* Y, *they necessarily agree in value of* Z, *for every tuple occurrence in* **R**
> FD: $Y \rightarrow Z$ if $\mathbf{r}^j(Y) = \mathbf{r}^k(Y)$ then $\mathbf{r}^j(Z) = \mathbf{r}^k(Z)$ *for all* $j, k \in I$.

Whenever $Y \rightarrow Z$ holds in a relation R (which usually reads "Y determines Z" or "Z is functionally dependent on Y") Y is called a determinant in R. Note that by Definition 3.2 every key is a determinant, that is all attributes are functionally dependent on every candidate key. In this sense, the notion of functional dependency is a generalization of keys.

Some functional dependencies are called trivial since they hold exactly for every relation (for instance, every attribute determines itself). The significance of trivial functional dependencies will become apparent when we consider normal forms.

Definition 6.2

> *A functional dependency* $Y \rightarrow Z$ *is said to be* ***trivial*** **if** $Z \subseteq Y$.

There are two important questions worth considering now. First, *how exactly do we determine functional dependencies in a given relation?* Second, *given a set of functional dependencies, how do we reckon that those specified are actually **all** dependencies that hold for the relation considered?*

Functional dependencies are the result of the data semantics, they reflect some rules or policies that are to be incorporated into the database in a form that is acceptable to the relational structure. Therefore, the only way of detecting functional dependencies is to analyse those rules and policies rather than, for instance, looking at particular instances of relations.

The answer to the second question is more complex. Details will be considered in Chapter 7; for now suffice it to say that for any given set of functional dependencies, say φ, it is possible to compute all such functional dependencies that are logically implied by φ. That is, given a set of rules we are able to detect all logically derivable functional dependencies (but not those of independent existence!).

Apart from a great theoretical significance to the whole of relational theory, functional dependencies can be used for practical problems of database design.

Example 6.3

Let COURSE-REGISTER be a relation holding details about all students taking some subjects throughout their course of study. A possible instance of this relation is depicted in Fig. 6.4.

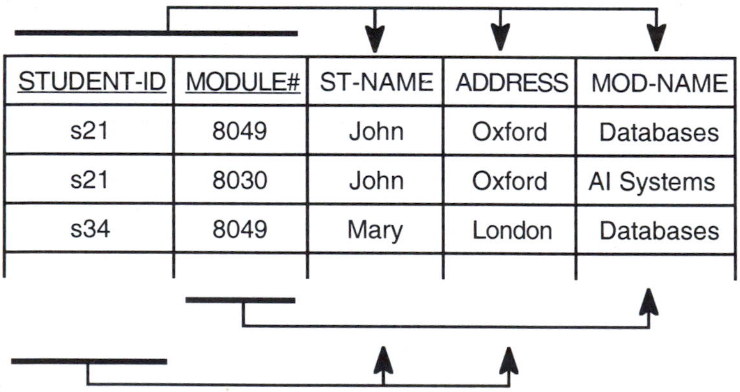

STUDENT-ID	MODULE#	ST-NAME	ADDRESS	MOD-NAME
s21	8049	John	Oxford	Databases
s21	8030	John	Oxford	AI Systems
s34	8049	Mary	London	Databases

Fig. 6.4

The primary key in this relation is a pair (Student#, Module#), hence it determines all the other attributes:

(Student#, Module#) → Student-Name
(Student#, Module#) → Address
(Student#, Module#) → Mod-Name

We can also identify three other functional dependencies:

Student# → Student-Name
Student# → Address
Module# → Mod-Name

These functional dependencies are graphically presented in Fig. 6.4 in the form of a so-called **dependency diagram.**

Now, we consider the relation COURSE-REGISTER in the context of certain possible cases of updating. We can easily notice that no student details (such as ST-NAME or ADDRESS) can be entered unless she or he attends some course. An attempt to insert a student who has not yet registered for any course, e.g. (S129, NULL, Pat, Oxford, NULL) would violate the principle of Entity Integrity and thus fail. For similar reasons, we could not include a new course, say (NULL, 8750, NULL, Compilers) until at least one student is registered for it.

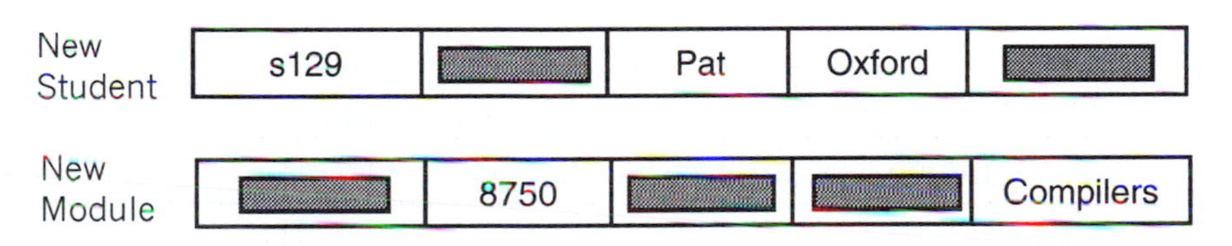

Relations with this kind of undesirable property are said to exhibit **insertion anomaly.**

COURSE-REGISTER misbehaves with respect to deletion, too. If the last student attending a particular module (say AI Systems in the example below) withdraws from it, we would loose all details about the course itself. This equally undesirable property is called **deletion anomaly.**

DELETION ANOMALY

Last Student	s21	8030	John	Oxford	AI Systems

Finally, modifying some values may prove cumbersome. If John moves from Oxford to another place, every tuple that refers to him would have to be updated and multiple updating always carries some risk of inconsistency. We call this kind of misbehaviour an **update anomaly**.

The above consideration leads us to a conclusion: normalized (1NF) relations may have undesirable update properties, hence bringing a relation to the first normal form would not terminate logical database design. Further transformations are needed to eliminate these kind of anomalies from a set of original relations.

Careful observation of COURSE-REGISTER permits us to identify the cause of the problem - some attributes are functionally dependent on a part of the key rather than the whole of it.

This brings us to the concept of the Second Normal Form (**2NF**). Loosely speaking, a relation is in 2NF if there are no partial functional dependencies.

Definition 6.3

A relation **R** *is in* **2NF** *if:*

- **R** *is in* **1NF** ***and***
- *every attribute of* **R** *is functionally dependent on the whole key but not any part of the key*

An immediate corollary from this definition is that any 1NF relation with a non-composite key is necessarily in 2NF.

To eliminate the harmful update anomalies in a given 1NF relation we need to decompose it into simpler ones using one of the allowable operations provided by Relational Algebra, namely *project*. We will shortly look at the problem of decomposition with more precision, but at this point let us consider this process on the following example.

Example 6.4

We consider the relation COURSE-REGISTER again. First, we decompose it by taking two projections:

STUDENT= **project** COURSE-REGISTER(STUDENT-ID,ST-NAME,ADDRESS)
T = **project** COURSE-REGISTER(STUDENT-ID, MODULE#, MOD-NAME)

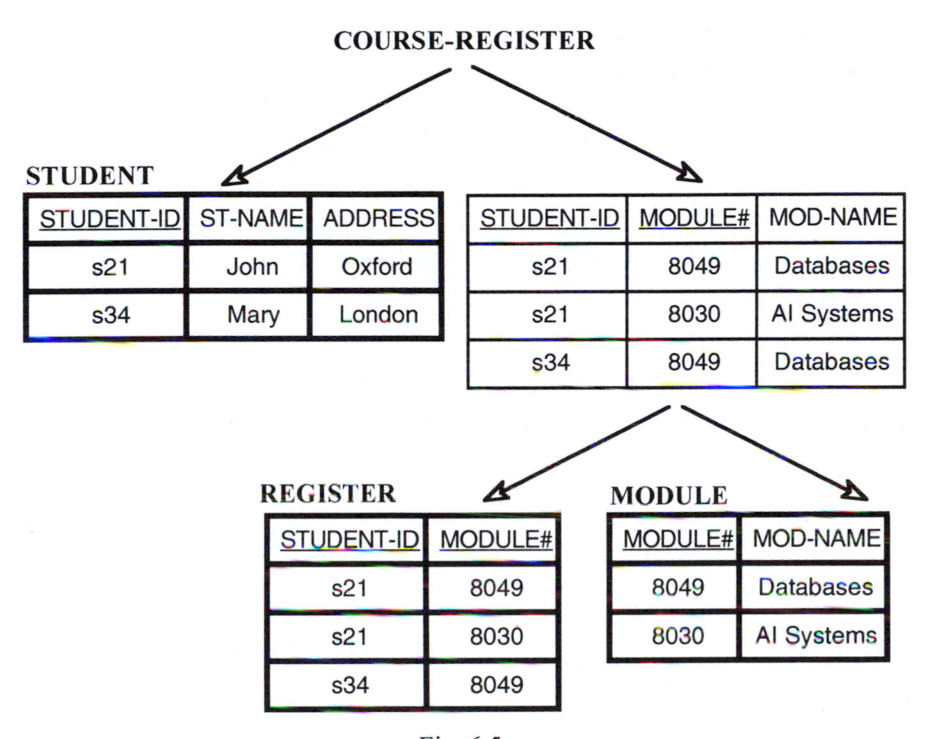

STUDENT

STUDENT-ID	ST-NAME	ADDRESS
s21	John	Oxford
s34	Mary	London

STUDENT-ID	MODULE#	MOD-NAME
s21	8049	Databases
s21	8030	AI Systems
s34	8049	Databases

REGISTER

STUDENT-ID	MODULE#
s21	8049
s21	8030
s34	8049

MODULE

MODULE#	MOD-NAME
8049	Databases
8030	AI Systems

Fig. 6.5

The relation STUDENT is now in 2NF - there are no partial dependencies here - though **T** is not; the functional dependency MODULE# $\rightarrow$ MOD-NAME still holds. Further decomposition of **T** by taking the following two projections:

REGISTER = **project** T(STUDENT-ID, MODULE#)
MODULE = **project** T(MODULE#, MOD-NAME)

will finally produce a family of relations (STUDENT, REGISTER, MODULE) that replaces the initial design. The reader might wish to verify that none of these three relations exhibit any of the earlier described update anomalies. In practical relational design 2NF is of relatively little use since the next higher

Boyce-Codd Normal Form (BCNF), which subsumes the former, can be achieved directly, i.e. without recourse to partial functional dependencies.

The above example suggests that in order to replace a set of initial relations by another set of improved (or better structured) relations we should decompose the initial ones by taking such projections as are implied by functional dependencies. This leads us, yet again, to the problem of representability. A more detailed treatment of this subject is given in Chapter 7; for now we need to appreciate in more general terms the requirement of decomposition without losing information.

We can formulate the problem as follows:

> *A set of relations* Σ *is equivalent to another set* Σ' *(that is* Σ *and* Σ' *represent the same information) if they contain the same attributes, and the associations amongst these attributes (such as functional dependencies) are satisfied in both representations.*

Some theoretical aspects of this problem are non-trivial. However, within the context of decomposition by projection this leads to a so-called property of *non-loss join*, a concept that can be understood quite well without subtleties of formal reasoning. This property can be formulated in the following way:

> *A decomposition of a relation* **R** *into its projections* $\mathbf{R'}_1, \mathbf{R'}_2, \ldots \mathbf{R'}_n$ *is* ***non-loss*** *if* **R** *can be recreated by joining these projections.*

Moreover, it can be proved that any relation **R**(X,Y,Z) that satisfies a functional dependency FD: X $\rightarrow$ Y can **always** be *non-loss* decomposed into its projections $\mathbf{R}_1$(X, Y) and $\mathbf{R}_2$(X, Z). For further considerations see Chapter 7 but the reader is urged to check (by the relevant tabulation) that the decomposition in Example 6.4 is indeed *non-loss*.

Some 2NF relation may still exhibit anomalies. We now consider an example to illustrate this point.

Example 6.5

Consider a relation that holds information about students and their tutors as shown in Fig. 6.6.

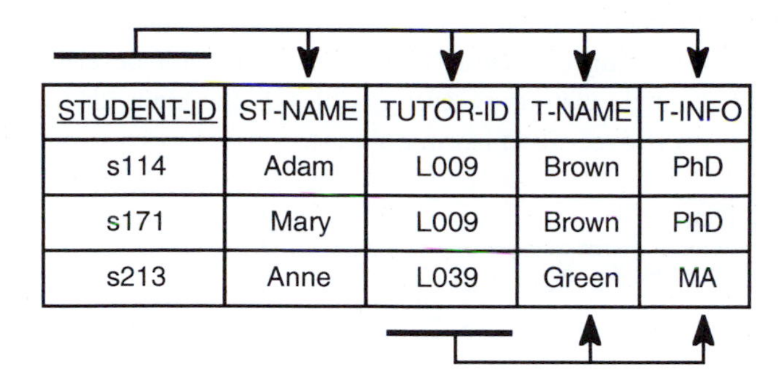

<u>STUDENT-ID</u>	ST-NAME	TUTOR-ID	T-NAME	T-INFO
s114	Adam	L009	Brown	PhD
s171	Mary	L009	Brown	PhD
s213	Anne	L039	Green	MA

Fig. 6.6

Here again the update anomalies occur. One cannot insert a tuple with some information about a new member of staff, say

		L231	Jones	Meng

who has not yet been assigned a personal student, since NULL value is not allowed for the primary key - STUDENT-ID. On the other hand, the deletion of a tuple that contains details of the last personal student of a particular member of staff (should a student decide to leave the college), e.g.

s213	Anne	L039	Green	MA

would also erase all the information about the relevant lecturer. The update anomaly also is there. For instance, if the value of T-INFO should change for a particular member of staff, it may be necessary to update many tuples. Clearly, the above relation contains too many concepts incorporated into its structure and should therefore be decomposed into two of its projections:

<u>STUDENT-ID</u>	ST-NAME	TUTOR-ID
s114	Adam	L009
s171	Mary	L009
s213	Anne	L039

<u>TUTOR-ID</u>	T-NAME	T-INFO
L009	Brown	PhD
L039	Green	MA

Fig. 6.7

It may be a useful exercise to check that the update anomalies have now been eliminated and that the above decomposition is indeed non-loss.

The misbehaviour demonstrated in Example 6.4 was actually the result of the initial relation not being conceptually simple. It did not describe one single concept about a single distinctive entity, which is the main idea of the strong third Normal Form (Boyce-Codd Normal Form).

Definition 6.4 — Boyce-Codd Normal Form

> *A* 1NF *relation* **R**(X_1, X_2, ...X_n) *is in* **BCNF** *if:*
>
> ***for every*** *attribute collection* χ *of* **R**
>
> ***if*** *any attribute not in* χ *is functionally dependent on* χ
>
> **then** *all attributes in* **R** *are functionally dependent on* χ

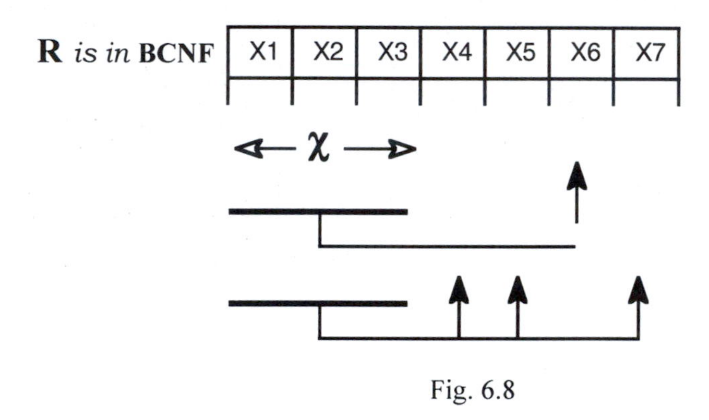

Fig. 6.8

An equivalent and most elegant definition of Boyce-Codd Normal Form is:

Definition 6.5

> *A relation is in* ***BCNF*** *if every determinant is a key*

The last definition can effectively be used for designing database relations. For a given relation, define functional dependencies and then check whether the key is the only determinant in this relation. If not, the appropriate non-loss decomposition yields (sub)relations that can again be verified and further decomposed until Definition 6.5 is satisfied. This simple algorithm is graphically represented in Fig. 6.9 below.

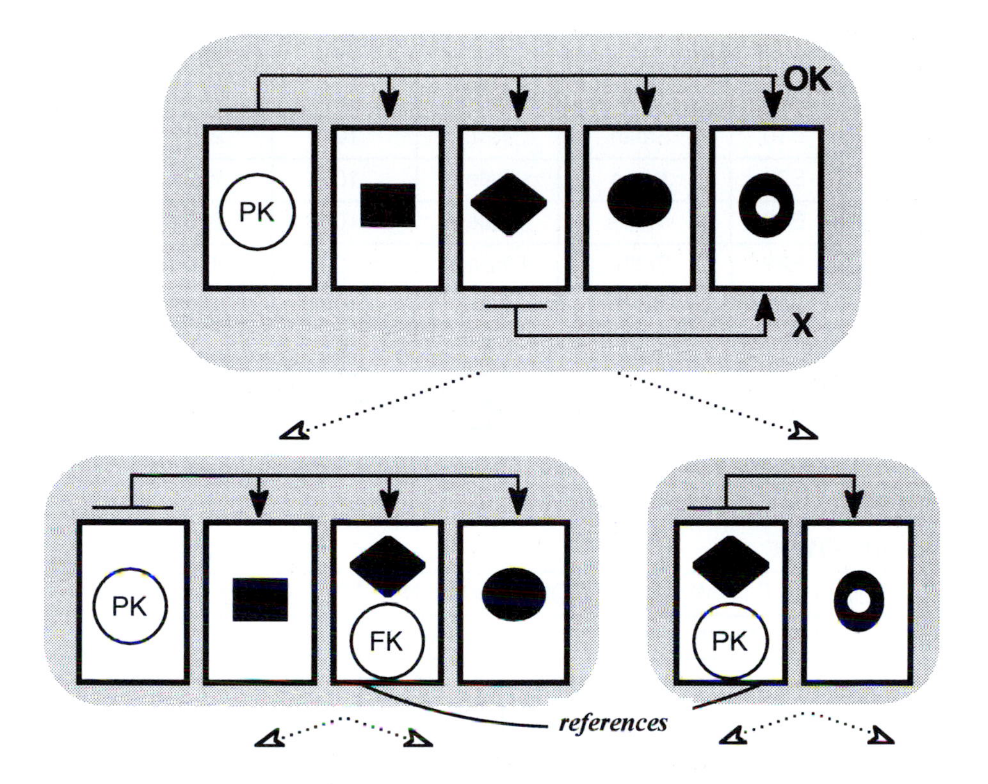

Fig. 6.9 BCNF decomposition schema

Problem 6.1

Consider a relation DIRECTORY whose possible instance is shown in Fig. 6.10. The following semantic rules are in force:

- Every employee works for some department but
- No employee works for more than one department
- No room is shared between departments
- Many employees may occupy one room
- Employee numbers are unique, employee names are not

Our task here is to normalize this relation to the BCNF.

DIRECTORY

E-NO	E-NAME	D-NAME	ROOM	TEL-NO
E10	Smith	R&D	101	2157
E21	Jones	Sales	105	2231
E18	Wilkes	Sales	105	2432
E22	Smith	Finance	108	4100
E09	Brown	Sales	109	2587

EMPLOYEE

E-NO	E-NAME	ROOM	TEL-NO
E10	Smith	101	2157
E21	Jones	105	2231
E18	Wilkes	105	2432
E22	Smith	108	4100
E09	Brown	109	2587

ALLOCATION

ROOM	D-NAME
101	R&D
105	Sales
108	Finance
109	Sales

Fig. 6.10

The semantic rules given at the beginning of this example define functional dependencies shown in a diagrammatical form. We notice that apart from all attributes being dependent on E-NO (the primary key), Room → D-NAME also holds for DIRECTORY. Thus ROOM is a determinant but not a candidate key. This violates the definition of BCNF and therefore DIRECTORY must be decomposed into two of its projections EMPLOYEE (containing details about every employee) and ALLOCATION which describes allocation of rooms to departments.

Problem 6.2

Consider a relation WORK-ALLOC whose possible instance is shown in Fig. 6.11. The following semantic rules are imposed:

E-NO	E-NAME	DEPT	ROLE	PROJECT	COST
E06	Smith	R&D	Programmer	Payroll	2
E06	Smith	R&D	Programmer	OODB	7
E02	Jones	Systems	Proj Leader	Payroll	2
E02	Jones	Systems	Consultant	OODB	7
E12	Smith	R&D	Analyst	Ledger	4
E06	Smith	R&D	Programmer	Ledger	4
E18	Green	Sales	Proj Leader	OODB	7

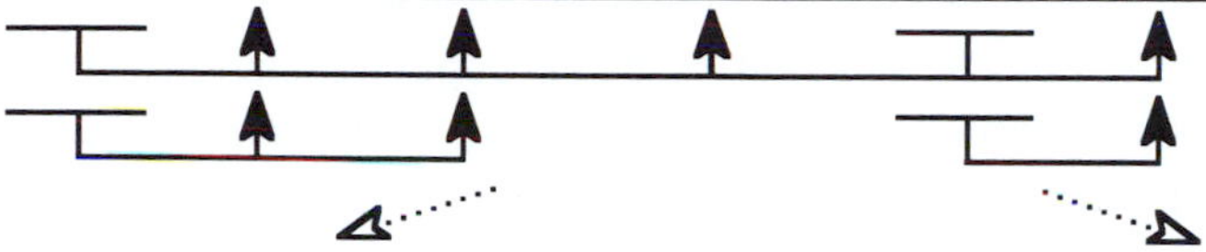

E-NO	E-NAME	DEPT	ROLE	PROJECT
E06	Smith	R&D	Programmer	Payroll
E06	Smith	R&D	Programmer	OODB
E02	Jones	Systems	Proj Leader	Payroll
E02	Jones	Systems	Consultant	OODB
E12	Smith	R&D	Analyst	Ledger
E06	Smith	R&D	Programmer	Ledger
E18	Green	Sales	Proj Leader	OODB

WORK

PROJECT	COST
Payroll	2
OODB	7
Ledger	4

EMPLOYEE

E-NO	E-NAME	DEPT
E06	Smith	R&D
E02	Jones	Systems
E12	Smith	R&D
E18	Green	Sales

ALLOC

E-NO	ROLE	PROJECT
E06	Programmer	Payroll
E06	Programmer	OODB
E02	Proj Leader	Payroll
E02	Consultant	OODB
E12	Analyst	Ledger
E06	Programmer	Ledger
E18	Proj Leader	OODB

Fig. 6.11

- Every employee works for some department but no employee works for more than one department
- An employee may work on many projects, possibly (though not necessarily) in different roles
- A team of one or more people carries out a project
- Employee numbers are unique, employee names are not.

The primary key in this relation is (E-NO, PROJECT) and therefore all other attributes are functionally dependent on this pair. However, three other functional dependencies:

E-NO → E-NAME, E-NO → DEPT, PROJECT → COST

also hold for WORK-ALLOC. Since none of these dependencies are defined on the whole key, the relation is not in BCNF (is not in the 2NF either) and needs to be decomposed.

Two subsequent non-loss decompositions are needed in order to obtain a family of three relations:

EMPLOYEE (E-NO, E-NAME, DEPT)
WORK (PROJECT, COST)
ALLOC (E-NO, ROLE, PROJECT)

each satisfying the necessary conditions to be in BCNF. In these relations the only functional dependencies are those resulting from the keys.

Case study

Normalize the relation SEMINAR exemplified in Fig. 6.12. The data in this relation are constrained by the following semantic rules:

(a) A seminar on a topic (such as *Modelling*) occurs only once on a date (eg. *20-05-97*); such an occurrence is given a unique identifier (SID). If there is another seminar on the same topic it will occur on a different date and will be given a separate identifier.

(b) A seminar is always on a topic and is given by a tutor in a specified room. At any given time only one seminar may occur in a room and only one tutor is present. The same topic may be taught by various tutors, however all of them will use the same book as a reference. A tutor may teach various topics but has a fixed, time-independent salary.

(c) A seminar is always related to a module (but not more than one). A module may have a number of seminars associated with it; the students are required to pass all of them in order to pass the module.

(d) Students get a single score on any seminar but may attend many seminars on the same topic. Only the highest score for a topic, however, counts towards the final result.

SEMINAR-RESULTS

SID	WHO	DATE	TUTOR	TOPIC	MOD	ROOM	MARK	BOOK	SAL
D29	John	20/5/97	Brown	EAR	8049	G117	71	Bowen	20
D29	Mary	20/5/97	Brown	EAR	8049	G117	86	Bowen	20
D17	John	28/2/97	Green	FD	8049	A109	98	Navathe	68
M09	Anne	18/4/97	White	Diff	8609	DG41	88	Kreitsky	55
D17	Mark	28/2/97	Green	FD	8049	A109	72	Navathe	68
U21	Sue	20/5/97	Brown	BCNF	8049	C101	91	Navathe	20
S31	John	22/3/97	Black	EAR	8049	DT21	73	Bowen	41
P43	Sue	30/5/97	Jones	Z	8043	C101	89	Hayes	29

Fig. 6.12

Clearly, to normalize a relation both identification of the primary key and the definition of functional dependencies are needed. The primary key must uniquely identify any tuple-occurrence; in the scenario given above each tuple is identifiable by the combination of the student-identifier and seminar-identifier since it refers to the result of a student on a seminar. Students are identified by their names (i.e. values of WHO) while seminars can be defined by any of the following:

SID → devised for that purpose
(DATE, TOPIC) → a seminar on a topic occurs only once on a date
(DATE, ROOM) → only one seminar may occur in a room at any time
(DATE, BOOK) → all tutors use the same book for a topic.

Hence the candidate keys are:

(WHO, SID)
(WHO, DATE, TOPIC)
(WHO, DATE, ROOM)
(WHO, DATE, BOOK)

Typically, the choice of primary key depends on volatility of its values (in principle, primary keys should not be updatable at all). In this case neither WHO nor SID needs to be updated and hence PK = (STUDENT, SID)

Functional dependencies need to be inferred from the semantics of the data given for this exercise in the forms of rules (a) through (d) above. Fig. 6.13 shows all of them together with a brief indication as to the reasons for which they occur:

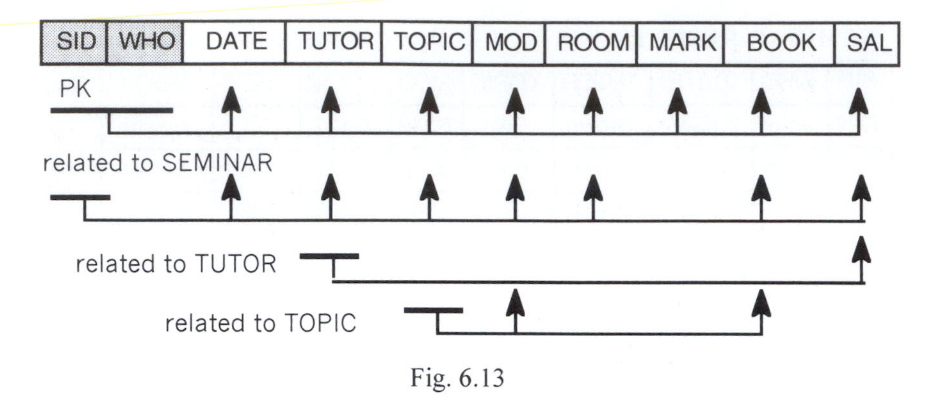

Fig. 6.13

Normalization of SEMINAR-RESULTS is presented graphically in Fig. 6.14. The final outcome of this process - BCNF relations - are marked with bold lines.

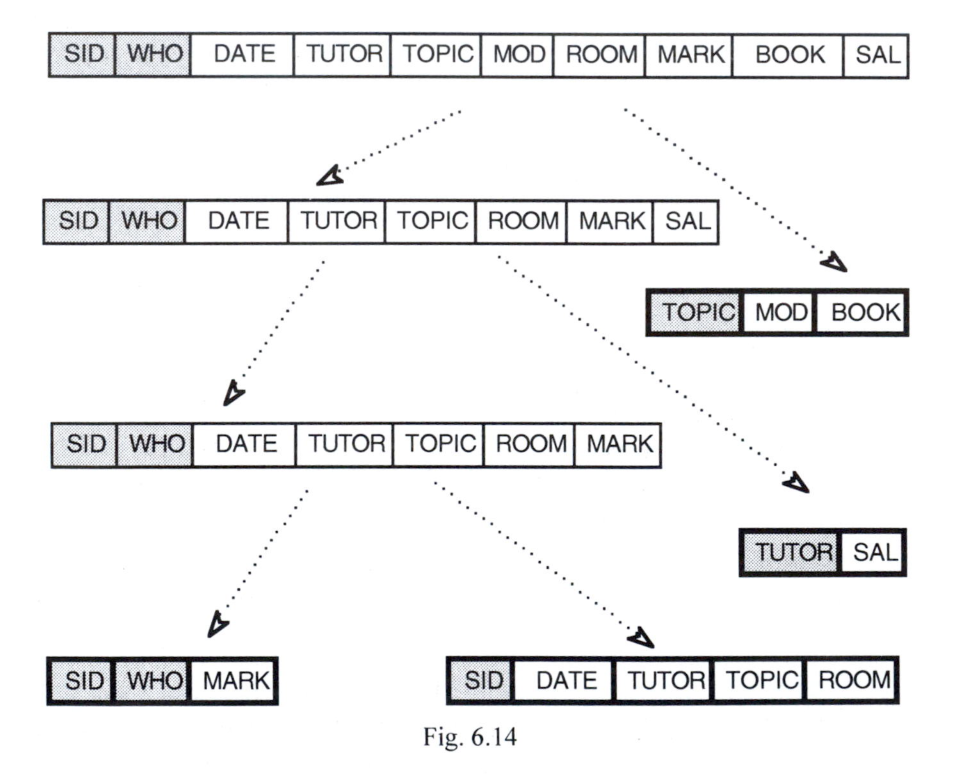

Fig. 6.14

Our database is then composed of four relations which interact in a way shown by means of Referential Integrity diagram in Fig. 6.15.

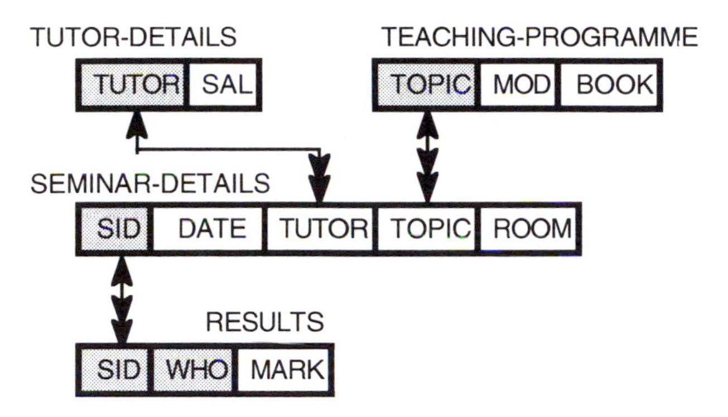

Fig. 6.15

Finally, an exercise on reverse engineering. The proper database design should commence with logical modelling, entities being then represented by relations which, in turn, are normalized thus extending the initial data model. The extent to which this happened can now be verified by deriving the EAR schema (shown in Fig. 6.16) that corresponds to BCNF relations:

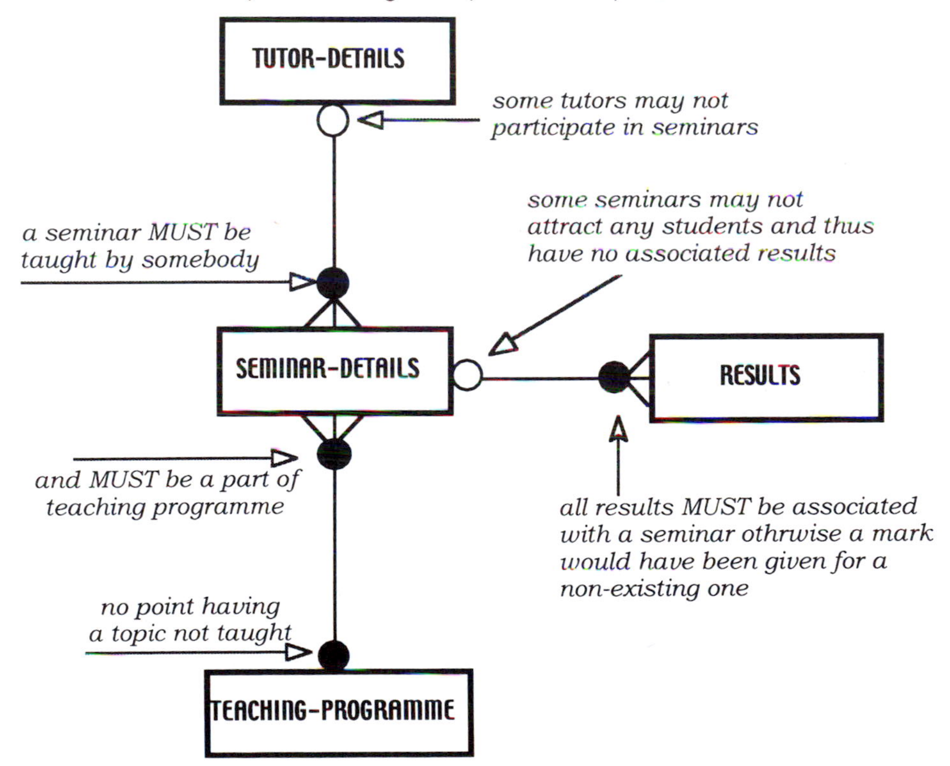

Fig. 6.16

Definition 7.2 **Fourth Normal Form**

> *A relation* **R**(X, Y, Z) *is in* ***4NF*** *if, whenever a non-trivial multivalued dependency* X→→Y *holds for R, then so does the functional dependency* X→A *for all attributes* A *in* **R**.

Ignoring certain formal subtleties, we can state an equivalent (and perhaps more practical) definition of 4NF:

Definition 7.3

> *A relation is in* **4NF** *if every multivalued dependency is also a functional dependency.*

As in the case of BCNF decomposition , definition 7.3 can effectively be used for normalizing database relations to the 4NF. For a given relation, it is necessary to check whether all MVDs are in fact FDs. If not, the appropriate non-loss decomposition yields (sub)relations that can again be verified and further decomposed until definition 7.3 is satisfied. Normalization to the 4NF would also guarantee all resulting relations to be in BCNF as the former is strictly stronger than the latter. Just as functional dependency is a generalization of keys, multivalued dependencies (MVD) can be seen as a kind of generalization of functional dependencies.

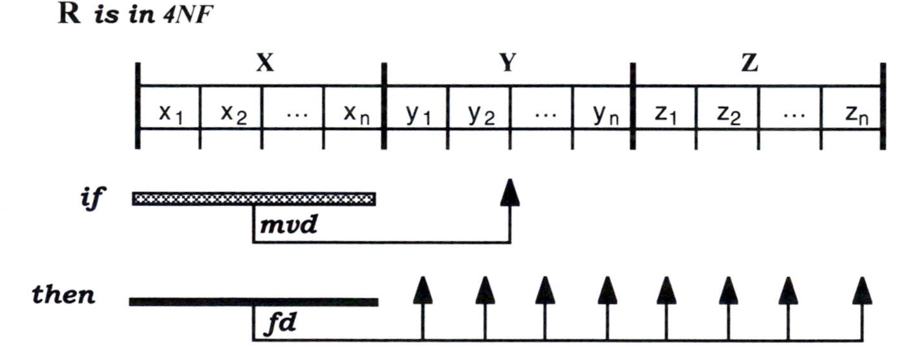

Fig. 7.3

Multivalued dependencies were discovered by Fagin (see Fagin 1977, on which this section is partly based) and independently by Zaniolo (1976). Their significance to the theory of relational databases is substantial, though their use in practical design processes is not particularly frequent. EAR modelling that normally precedes the normalization phase tends to lead to a reasonably well-formed relational representation where no need for removing logical flaws arises.

To complete Example 7.1: since Student→→Course and Student→→Hobby hold in the original relation and neither of them is a functional dependency, we have to decompose the original relation into its two projections as shown in Fig. 7.4.

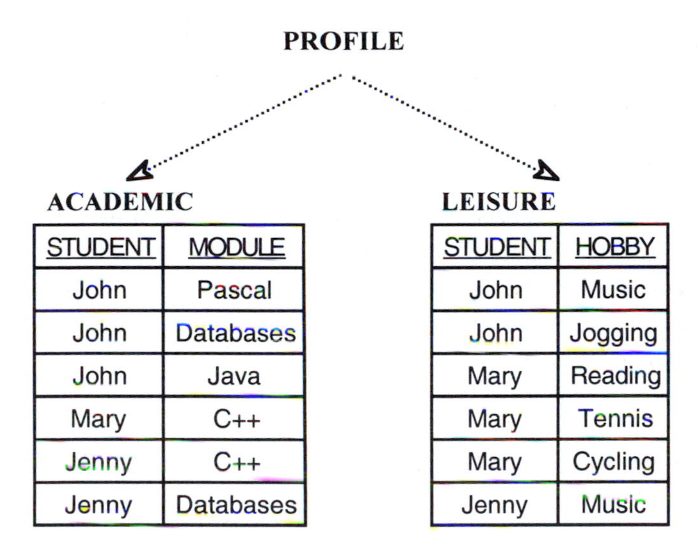

ACADEMIC

STUDENT	MODULE
John	Pascal
John	Databases
John	Java
Mary	C++
Jenny	C++
Jenny	Databases

LEISURE

STUDENT	HOBBY
John	Music
John	Jogging
Mary	Reading
Mary	Tennis
Mary	Cycling
Jenny	Music

Fig. 7.4

7.2 JOIN DEPENDENCY. FIFTH NORMAL FORM

Multivalued dependencies are based on the notion of mutually independent intra-relationships between attributes in a table. Sometimes these relationships may be constrained further by the imposition of (somewhat erroneous from a pure logic viewpoint) cyclic propositions; hence the 4NF structures would, yet again, have undesirable update properties. Let us consider an example to illustrate the point.

Example 7.2

Consider relation **AWARD** that holds some data about UNIVERSITIES awarding DEGREES in some DISCIPLINES (see Fig. 7.5).

AWARD

UNIVERSITY	DISCIPLINE	DEGREE
Old Town	Computing	BSc
Old Town	Mathematics	PhD
New City	Computing	PhD
Old Town	Computing	PhD

Fig 7.5

This somewhat redundant form is necessary since the relation is to represent the following fact: UNIVERSITY teaches discipline in which students can get a DEGREE. We have, in fact, defined three propositions on the attributes of the relation **AWARD**:

teaches (UNIVERSITY, DISCIPLINE)
is_read_for (DISCIPLINE, DEGREE)
awards (UNIVERSITY, DEGREE)

Due to their cyclic nature, these propositions require **AWARD** to be in the form as given above. For example, although Old Town University teaches Mathematics and awards BSc , BSc in Mathematics was not necessarily awarded, i.e.:

teaches (Old Town, Mathematics) = true
supplies (Old Town, BSc) = true
is_read_for (Mathematics, BSc) = false

An interesting consequence of the cyclic propositions imposed on a relation structure is that the relation cannot be non-loss decomposed into **two** of its projections; it can though be decomposed into **three** of its projections, as illustrated in Fig. 7.6.

Notice that joining any two projections will produce a result that contains an extra 'spurious' tuple, that is a tuple which does not exist in the original relation. Further join will eliminate this false information though. The reader may wish to verify that this happens irrespective of the order in which the three projections are re-joined.

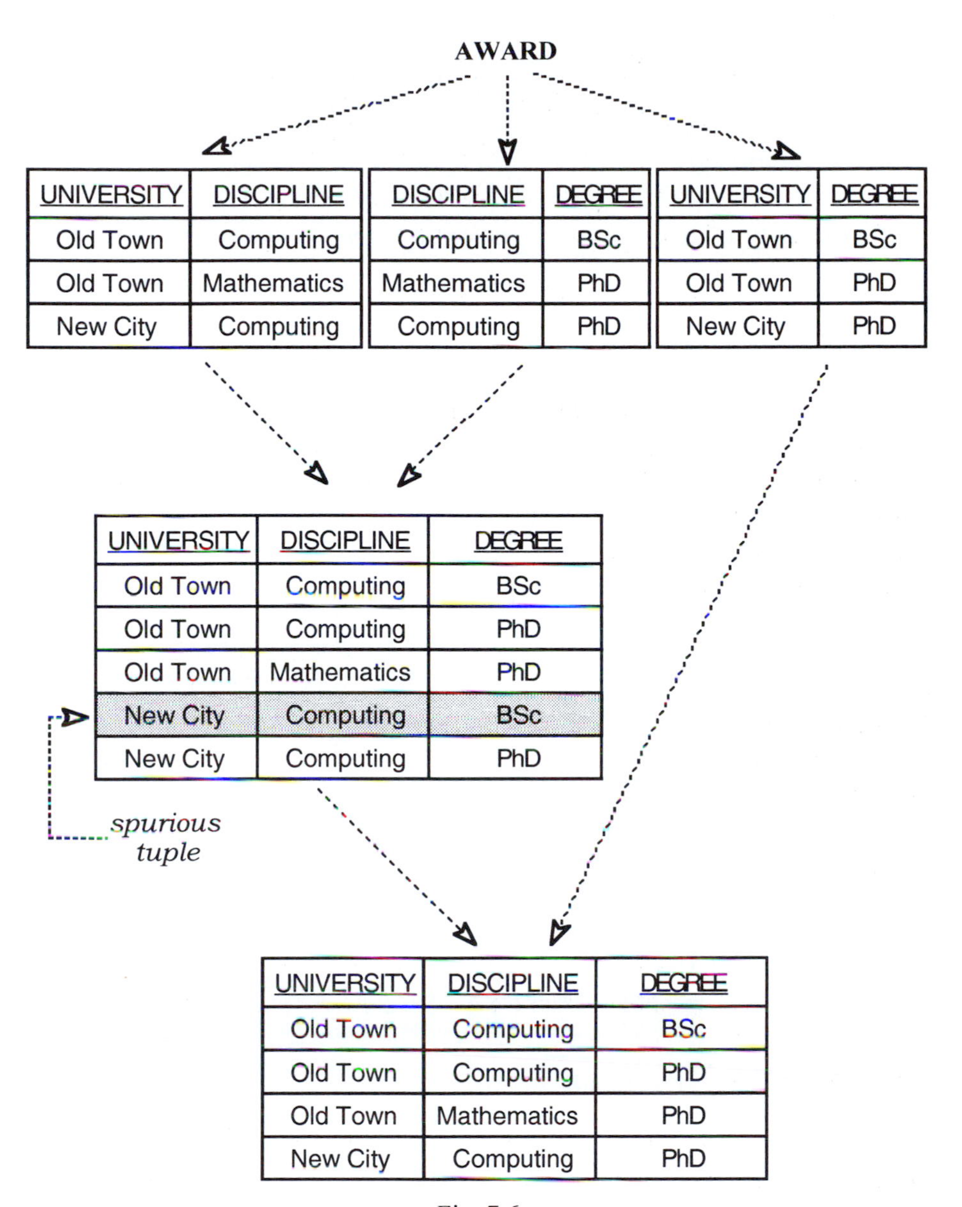

UNIVERSITY	DISCIPLINE
Old Town	Computing
Old Town	Mathematics
New City	Computing

DISCIPLINE	DEGREE
Computing	BSc
Mathematics	PhD
Computing	PhD

UNIVERSITY	DEGREE
Old Town	BSc
Old Town	PhD
New City	PhD

UNIVERSITY	DISCIPLINE	DEGREE
Old Town	Computing	BSc
Old Town	Computing	PhD
Old Town	Mathematics	PhD
New City	Computing	BSc
New City	Computing	PhD

UNIVERSITY	DISCIPLINE	DEGREE
Old Town	Computing	BSc
Old Town	Computing	PhD
Old Town	Mathematics	PhD
New City	Computing	PhD

Fig. 7.6

Definition 7.4 **Join Dependency**

> *Let* **R** *be a relation and* $\mathbf{R_1}, \mathbf{R_2}, \mathbf{R_3}, ..., \mathbf{R_m}$ *be a set of projections that constitute a non-loss decomposition of* **R**.
> *Join dependency*
> $JD^* (\mathbf{R_1}, \mathbf{R_2}, \mathbf{R_3}, ..., \mathbf{R_m})$ *holds in* **R**
> ***if and only if*** **R** *is equal to the join of these projections.*

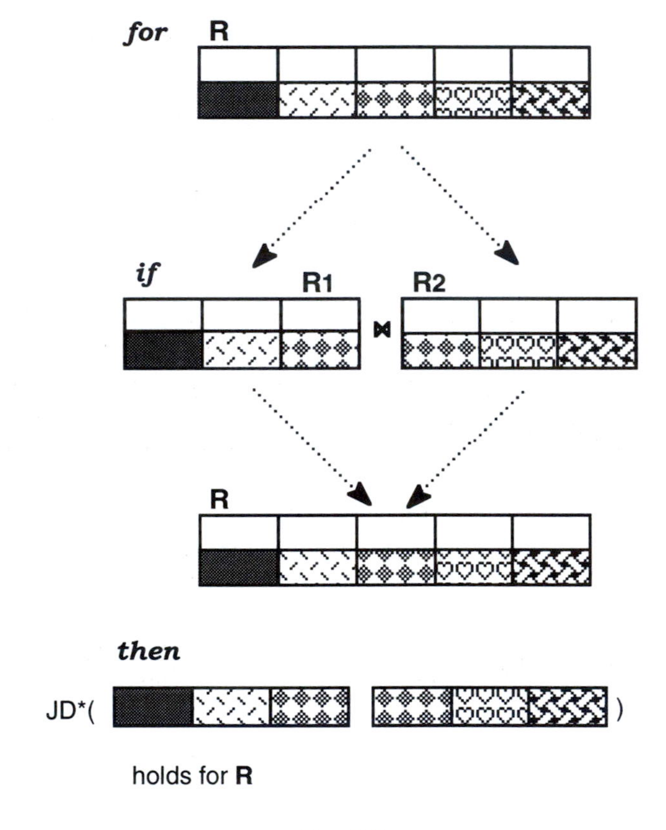

Fig. 7.7

The immediate corollary is that every multivalued dependency is also a join dependency. Join dependency is then a further generalization of the concept of intra-relationships between attributes within one relation. It is expressed as a conjunction of a number of predicates defined on attributes, and in this sense can be regarded as the most general kind of constraint that could be imposed on the relational structure with respect to the pair of the algebraic operations *project* and *join.*

In Example 7.2, the relation **AWARD** satisfies the join dependency

JD * ((UNIVERSITY, DISCIPLINE),
(DISCIPLINE, DEGREE),
(UNIVERSITY, DEGREE))

and (as the reader may wish to verify) is not equivalent to any collection of multivalued dependencies

UNIVERSITY→→DISCIPLINE	DISCIPLINE→→DEGREE,
UNIVERSITY→→DEGREE	DISCIPLINE→→UNIVERSITY,
DEGREE→→DISCIPLINE	DEGREE→→UNIVERSITY.

Join dependency is called trivial (since it holds for exactly every relation) if one of the projections $\mathbf{R}_i$ is the relation **R** itself.

Definition 7.5 — Fifth Normal Form

A relation **R** *is in* **5NF iff** *for all non-trivial join dependencies* JD * ($\mathbf{R}_1, \mathbf{R}_2, \mathbf{R}_3, ..., \mathbf{R}_m$) *that hold for* **R**, *every projection* $\mathbf{R}_i$ *is a superkey for* **R**.

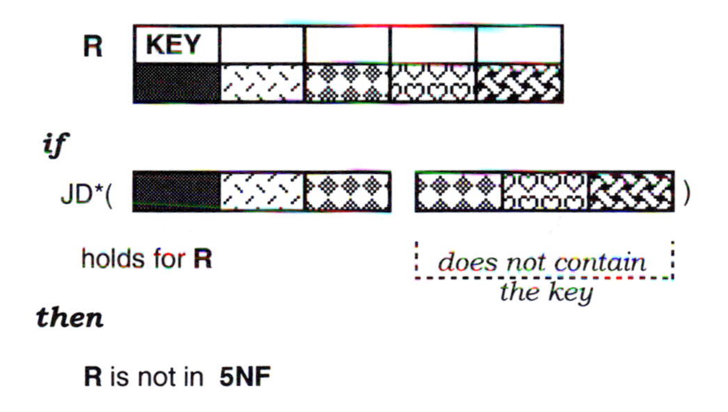

Fig. 7.8

Because every multivalued dependency is also a join dependency, every 5NF relation is also in 4NF. It is possible to prove (Fagin, 1979) that any given relation can be non-loss decomposed into an equivalent collection of 5NF relations.

Reflexivity	***If*** $Y \subseteq X$ ***then*** $X \rightarrow Y$, *for* $X, Y \subseteq \mathbf{U}$

It is relatively straightforward to verify that in any relation a set of attributes functionally determines all its subsets; if Y = (A, B, C) then the following dependencies hold: Y → A | B | C | (A, B) | (A, C) | (B, C) | (A, B, C)

The reflexivity rule only generates trivial dependencies, and thus finds its use in formal derivation of logical formulae that involve functional dependencies.

Augmentation	***If*** $X \rightarrow Y$ ***then*** $(Z, X) \rightarrow (Z, Y)$ *for* $X, Y, Z \subseteq \mathbf{U}$

As an example, consider a relation **RESIDENCE** whose possible instance is shown below:

RESIDENCE

STUDENT	HALL	ROOM	TYPE	RENT
John	Morrell	101	Double	60
Dave	Morrell	101	Double	60
Anne	Warnford	305	Single	70

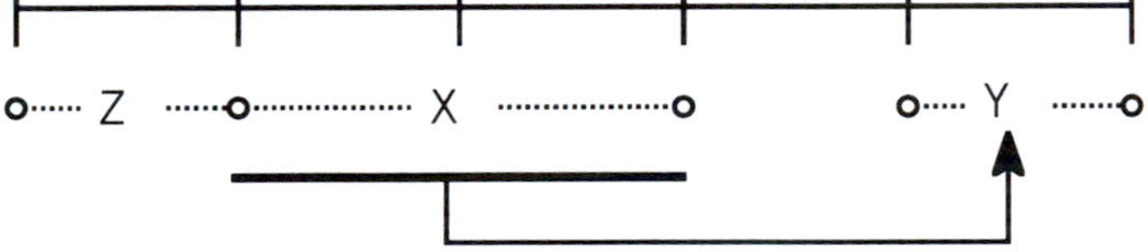

Z X Y

Fig. 7. 10

Since (HALL, ROOM) → RENT and STUDENT is an attribute of **RESIDENCE** then, with respect to the augmentation rule, (STUDENT, HALL, ROOM) → (STUDENT, RENT) also holds.

Transitivity	***If*** $X \rightarrow Y$ ***and*** $Y \rightarrow Z$ ***then*** $X \rightarrow Z$ for $X, Y, Z \subseteq \mathbf{U}$

Consider a relation ENROLMENT whose possible instance is shown in Fig. 7.11. Should we assume that MODULE → LECTURER (at most one lecturer teaches a module) and LECTURER → DEPARTMENT (no lecturer teaches for more than one department) then - with respect to the transitivity rule - the functional dependency MODULE → DEPARTMENT holds, too.

ENROLMENT

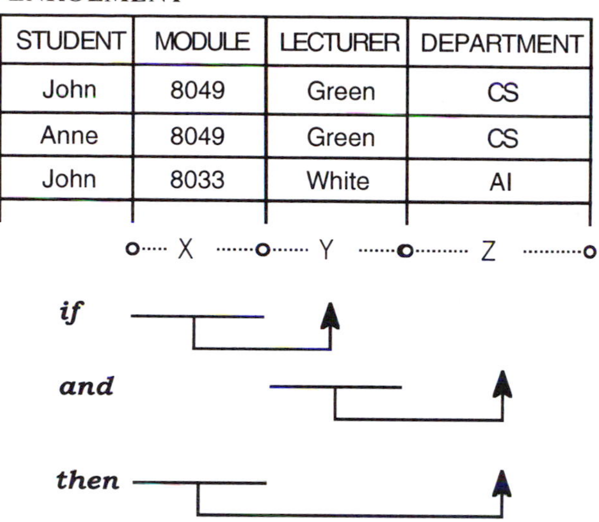

STUDENT	MODULE	LECTURER	DEPARTMENT
John	8049	Green	CS
Anne	8049	Green	CS
John	8033	White	AI

Fig. 7.11

The Armstrong axioms are **sound** and **complete**. They are sound because they do not generate any incorrect dependencies and complete since **all** functional dependencies implied by φ can be derived from φ using these axioms.

Two further rules can be derived from the Armstrong axioms and offer further help in computing the closure φ^+. The first one is known as the union/decomposition rule, whereas the second is referred to as the pseudo-transitivity rule.

Lemma 7.1

($X \rightarrow Y$ ***and*** $X \rightarrow Z$) ***if and only if*** $X \rightarrow YZ$.

Proof (for notational simplicity, we assume $(X, Y) \equiv XY$).

Left-implication

Union	***if*** ($X \rightarrow Y$ ***and*** $X \rightarrow Z$) ***then*** $X \rightarrow YZ$

By augmentation

Example 7.7

Consider a relation SKILL. Assuming that student names are unique within the class, the primary key in SKILL is (STUDENT, TYPE). A possible instance of this relation is given in Fig. 7.19 where functional dependencies are shown diagrammatically.

SKILL

<u>STUDENT</u>	<u>TYPE</u>	LANGUAGE
John	Web Site	Java
Mark	Web Site	Pascal
John	Database	SQL
Anne	Web Site	Java
Mark	Database	SQL

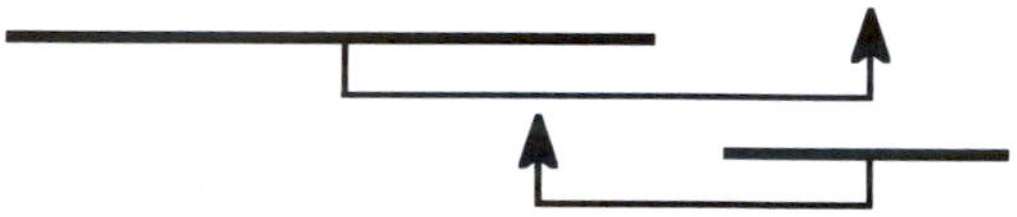

Fig. 7.19

It's easy to verify that SKILL conforms to the definition of third normal form (as the only non-prim attribute LANGUAGE is not transitively dependent on the primary key). Equally, SKILL is not in BCNF since LANGUAGE is a determinant but not a candidate key - which we can verify by computing the closure of LANGUAGE under given functional dependencies.

Is LANGUAGE a superkey for SKILL?			
aux	fd: $\alpha \to \beta$	$\alpha \subseteq$ aux ?	aux'
L	**ST → L**	**no**	**L**
L	**L → T**	**yes**	**LT**
LT	**ST → L**	**no**	**LTL = LT**
LT	**L → T**	**yes**	**LTT = LT**
L^+ := LT - not the same as SKILL = STL			

Thus to achieve BCNF further decomposition is necessary as shown in Fig. 7.20.

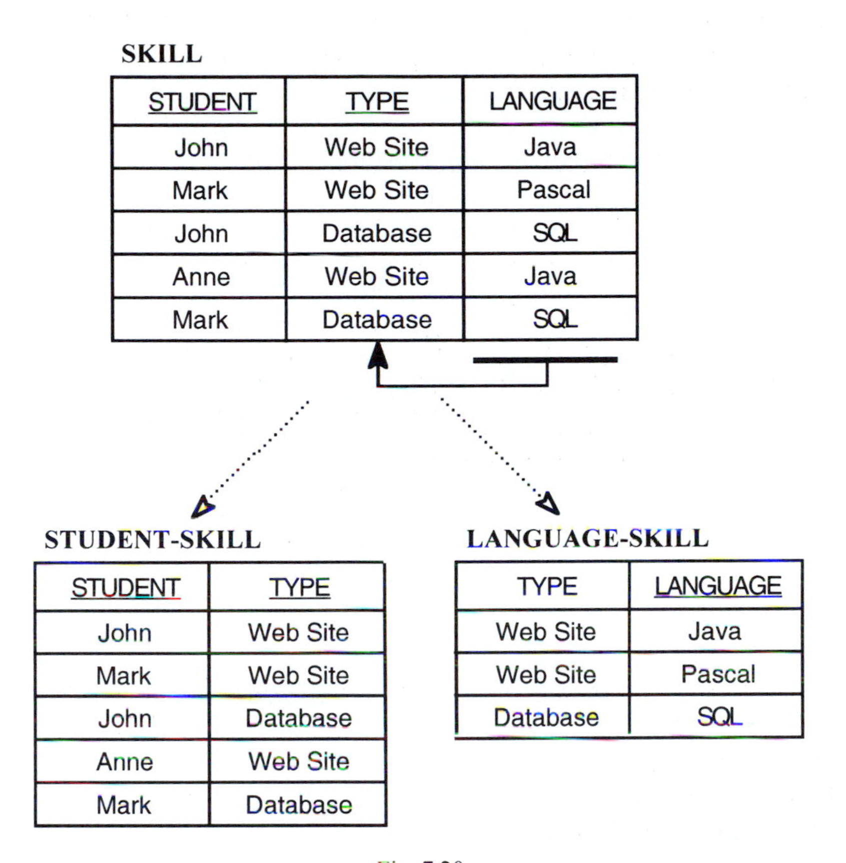

SKILL

STUDENT	TYPE	LANGUAGE
John	Web Site	Java
Mark	Web Site	Pascal
John	Database	SQL
Anne	Web Site	Java
Mark	Database	SQL

STUDENT-SKILL

STUDENT	TYPE
John	Web Site
Mark	Web Site
John	Database
Anne	Web Site
Mark	Database

LANGUAGE-SKILL

TYPE	LANGUAGE
Web Site	Java
Web Site	Pascal
Database	SQL

Fig. 7.20

The purpose of this example is to demonstrate that non-loss BCNF decomposition, although always possible, may not preserve functional dependencies. On the other hand, 3NF is always achievable with functional dependencies intact. Hence, if dependency preservation is more important for a given application system than its complete updatabillity, 3NF rather than BCNF might be preferable. In either case the maintenance procedures (i.e. those handling insertion, deletion, modification) will need appropriate adjustments to prevent violation of referential integrity.

So far we have assumed that any relation can be decomposed (through using *project*) without any loss of information (and hence recomposed via *join*). We now prove that this property universally holds. The theorem was originally formulated in Heath (1971) with an improved proof given in Rissanen (1979). Here we follow a proof given in Ullman (1982) with some minor modifications.

Heath Theorem

> *A relation* **R**(X,Y,Z) *that satisfies a functional dependency* X → Y *can always be non-loss decomposed into its projections* $\mathbf{R}_1$(X, Y) *and* $\mathbf{R}_2$(X, Z).

In what follows we are going to use the following denotations:

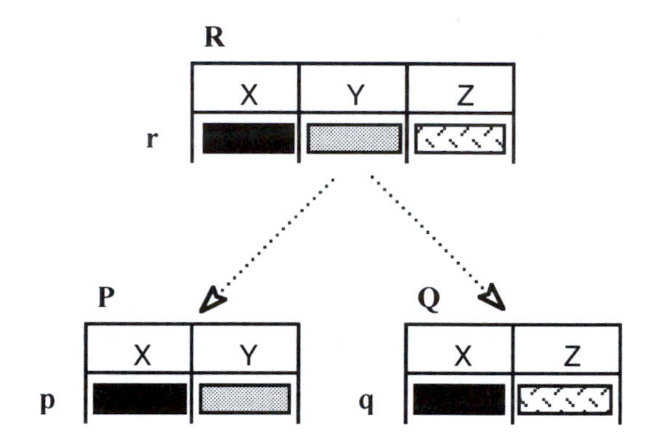

Lemma 7.3

> *If* **P** = **project R**(X, Y) *and* **Q** = **project R**(X, Z)
> *then* **R** ⊆ **join (P, Q**: [**P**.X = **Q**.X])

Proof

Let **r** be a tuple from **R**. Then, there is a tuple from **p** in **P** such that:
r(X, Y) = **p**(X, Y)

and there is a tuple **q** in **Q** such that:
r(X, Z) = **q**(X, Z)

By definition of the natural join
r is in **join** (**P**, **Q**: [**P**.X = **Q**.X])
since **r**(X) = **p**(X) = **q**(X)

QED

Lemma 7.4

> *If* **S** = **join** (**P**, **Q**: [**P**.X = **Q**.X])
> *then* **P** = **project S**(X, Y) *and* **Q** = **project S**(X, Z)

Proof

By Lemma 7.3 **R** ⊆ **S** and therefore
project R(X, Y) ⊆ **project S**(X, Y)
project R(X, Z) ⊆ **project S**(X, Z)

Consequently,
P ⊆ **project S**(X, Y)
Q ⊆ **project S**(X, Y)

We need to prove that **P** = **project S**(X, Y) and **Q** = **project S**(X, Z) which can be done by showing that :
project S(X, Y) ⊆ **P** and
project S(X, Z) ⊆ **Q**.

Suppose a tuple **p** from **P** is in **project S**(X, Y). Then there is a tuple **s** in **S** such that **s**(X, Y) = **p**(X, Y). Consequently, there are such tuples **p**' and **q**' that **s**(X, Y) = **p**' and **s**(X, Z) = **q**'. Thus **s**(X, Y) is in **P** and **s**(X, Z) is in **Q**. Since **s**(X, Y) = **p** it follows that **p** is in **P**.

QED

7.5 EXERCISES

7.1 Verify the following definition of functional dependency:
X → Y holds for **R** = (.., X, Y, ..) **iff**
for all x ∈ X, cardinality(**ΠY σ** (R: X=x) ≤ 1

7.2 Outline proofs of the following theorems:
(a) *every BCNF relation is necessarily in 2NF*
(b) *every 4NF relation is necessarily in BCNF*
(c) *BCNF is strictly stronger than 3NF*
(d) *every 5NF relation is necessarily in 4NF.*

CHAPTER 8

Structured Query Language

8.1 INTRODUCTION

The relational model comprises a Data Sub-Language (DSL) whose purpose is to retrieve and to modify the data. The most commonly known basis for relational DSL is Relational Algebra. However, algebraic operations are defined in a non-programming fashion (essentially, they are re-defined set-theoretical operators, for the basic data structure is some kind of set) and are based on the principle of relational closure. These operators were not necessarily meant to be directly implemented, yet two important concepts assumed for Relational Algebra were retained for all subsequent programming implementations. First, the operands are the whole sets of similarly structured records rather than individual records. Second, the non-procedural character of expressions relieves the programmer from specifying the details of *how* (in the sense of algorithmic logic) the results should be obtained.

The first attempt to implement a language based on the principles of the relational model was undertaken in the IBM Research Laboratory (San Jose) as a part of the experimental work on System R. SEQUEL (*Structured English Query Language*), subsequently renamed SQL (*Structured Query Language*), was developed and supplemented by additional facilities usually required in file processing. Thus, apart from the relational capabilities, the language incorporated some structures that allow programmers to define the permissible data structures, to carry out computations (arithmetic functions, date functions, extended logical functions), to perform string processing, to format reports and many others.

SQL has become a standard relational language, ratified by ANSI in 1986 and by ISO in 1987. Some very important features were not, however, properly specified and, as a consequence, many if not most of the then marketed software packages have not supported in a sufficient manner such concepts as domains, primary keys and foreign keys.

This situation was partly rectified later by incorporating *Integrity Enhancements* into the standardizing documents (ISO/ANSI 1989). Three years later SQL 2 standard (incorporating program structures) was approved and SQL 3 (adding object -relational facilities) was released in 1999 (and re-named SQL-99).

This rather long (and continuing) standardization history reflects the fact that SQL was originally perceived as a **database language** rather than a **programming language** and therefore its standard version did not support typical program structures such as *if...then...else*, *while...do*, *procedure* at all. SQL expressions were supposed to be either short interactive (interpreted) queries or external procedures to be called from within a program written in a *host* programming language (mostly C) that would have full operational facilities.

At present there are some hundred implementations of SQL, and although they differ (amongst themselves and from the standard), the fundamental concepts of relational processing are retained across the whole range.

As far as possible, the SQL expressions given in this Chapter are constructed on the basis of the standard (basic) version of the language. However, we found it rather difficult to restrict this text to the standard version only. Thus, a number of implementation-dependent operations (for instance, compound transactions) are expressed in ORACLE*SQL due to its relative independence of any particular hardware.

8.2 DEFINING DATABASE OBJECTS

SQL is a language that allows the programmer to formulate expressions (i.e. programs) that represent various processes for database interrogation or updating or restructuring. These expressions are typically interpreted (rather than compiled) and executed instantaneously. However, before any manipulation of the data can be done, the necessary structures to hold that data must be defined.

The only data structure capable of holding the data is a **table**. The word *table* rather than *relation* is used here since, in principle, any table can be defined, and this includes unnormalized and badly structured tables. SQL does not provide any mechanism to detect or correct any structural improprieties. A table consists of named **attributes** (columns). For each attribute its type is specified (Fig. 8.1); more complex semantic definitions (e.g. restrictions of values, interrelationships among attributes) are supported via user-defined **constraints**.

char (n)	Fixed length string of length **n** bytes. Max **n** = 255 B, default **n** = 1 B
varchar2(n)	Currently synonymous with **varchar(n)**. Variable length string of length **n** bytes. Max **n** = 2000 B
number (length, fraction)	**length** = total number of digits {1, 38} **fraction** = length of fraction part {-84, 127} permissible magnitude $\{10^{-130}, 10^{126} -1\}$
number (length)	fixed point number (no fraction part)
number = float	floating point number with length = 38
date	range {01-Jan - 4712 BC, 31-Dec 4712 AD} default format 'dd-mmm-yy'
long	variable length string, max $2^{31} -1$ B one column per table, non-indexable, search-inactive, no integrity constraints can be defined on (except **not null**)

Fig. 8.1 SQL data types (Oracle)

Three kinds of table are typically supported by a DBMS. The **base tables** are those that correspond to the relations devised in the conceptual schema. They physically exist (albeit not in the form of the simple mono-type sequential files) and, unless updated or restructured, remain relatively static throughout the life-time of a database system. The **derived tables** are obtainable from the base tables. They contain some results of querying and therefore their normal life time extends only over the period of query execution. If required, they can be stored permanently by explicitly writing the result of a query into a predefined table, rather than (implicitly) directing it into the assigned workspace.

SQL uses extensively the concept of a **virtual table** (view). A view can be thought of as a portion of the database that is of a specific interest to a particular group of users (or processes). A view is normally defined on one or more base tables, its content being restricted by a set of conditions. To the user a view looks exactly like a permanently stored table while the system maintains it physically via the relevant system of pointers.

Any table that is permanently stored in the database can be given an alternate name, i.e. a synonym. The reasons for using synonyms are two-fold:

> ***Convenience***. Sometimes the original table names are rather long (for they are supposed to carry some meaning); using the appropriate shorthand instead makes typing the SQL expression less tedious.
> ***Expression Integrity***. Formulating unambiguously certain queries would not be possible (joining a table with itself is a prime example since this operation carries a risk of table corruption) without using a substitute name for a table.

SQL enforces Entity Integrity through specification of **constraints**. A constraint set on the prime attribute(s) ensures both the uniqueness and existence of the primary key - any attempt to insert a duplicate value for the primary key or to insert a record with a **null** value for it will necessarily fail. Furthermore, any attribute in a table can be restricted (if needed) not to accept null values.

In the current version of SQL the principle of Referential Integrity is only partly supported. A mechanism is provided to inform the DBMS about the existence of the foreign keys and their correspondence to the relevant primary keys. The constraint definition can also inform the DBMS whether or not deletion of referenced records is automatically required should the corresponding primary key be deleted (*cascade deletion*). Provision for maintaining referential integrity with respect to insertion is much weaker - records containing a foreign-key value that does not match the corresponding primary key will not be inserted.

Apart from general integrity constraints, the application-dependent particular integrity rules can be enforced through **check**-constraints that explicitly define conditions which must be obeyed by all the tuples in the table. The check-constraint can refer to any attribute in the table but cannot refer to the attributes of other tables.

A non-unique **index** can also be defined on any attribute (or a combination of attributes). Such a definition will logically invert the file that stores the relevant table with respect to the specified attributes. While this may speed up certain retrievals it also produces a substantial overhead.

From a programming viewpoint, SQL belongs to imperative languages. An SQL statement is effectively a command, though not as in the conventional languages such as Pascal, C or Java. SQL is genuinely orientated towards *set-at-a-time* rather than *record-at-a-time* processing. In this context the lack of support for the relational assignment (i.e. a construct like R := Q in the conventional languages except that R and Q would denote relations) is especially annoying though it can be overcome by using the same technique as for converting the derived tables into base tables.

To define a database object the **create**-statement is used. Its simplified syntactic form is depicted in Fig. 8.2 . The create statement is meant to be dynamic and, in principle, can be used at any time - as and when required.

```
create table table-name
    (attribute-name data-type
    [default expression]
    [constraint constraint-name                    (* column constraint *)
        {primary key | unique |
        references table-name(attribute-name) [on delete cascade] |
        not null | check condition}],
    (attribute-name data-type
    [default expression]
    [constraint constraint-name                    (* column constraint *)
        {primary key | unique |
        references table-name(attribute-name) [on delete cascade] |
        not null | check condition}],
        ......................................,
    [constraint constraint-name                    (* table constraint *)
        {primary key(attribute-names) |
        unique (attribute-names) |
        foreign key (attribute-names) references
            table-name(attribute-names) [on delete cascade] |
            check condition}]
    );
```

Fig. 8.2 Various syntactic forms of the *create table* statement

create view as query **with check option constraint** constr-name
create index-name **on** table-name (attribute-name, attribute-name, ...)
create sequence -name **increment by** integer **start with** integer
create [**public**] **synonym for** {table-name | view-name | sequence | synonym}

Fig. 8.3 Creating database objects

The object thus created is immediately available and will be maintained by the DBMS until a command to destroy it is issued. Fig. 8.4 shows the form of the *drop*-command.

```
drop
        {table table-name  cascade constraints |
        view view-name |
        index index-name |
        sequence sequence-name |
        [public] synonym synonym-name}
```

Fig. 8.4 Simplified syntactic definition of the *drop*-statement

The effect of dropping an object from the database is also immediate and irrevocable. In the case of a table it is also independent of whether or not the table contains some data. Hence, for instance the statement: **drop table** MYTABLE will destroy the table structure **and** erase its content - whatever it might be!

The table definition can be changed by using the *alter*-statement. This again is a dynamic instruction and can be used at any time with an immediate effect.

```
alter table table-name
      { add (attribute specification as in the create statement) |
      modify attribute-name [data-type] [default expression] [column-
      constraint] |
      drop
          {primary key [cascade] |
          unique (attribute-names) |
          constraint constraint-name}}
```

Fig. 8.5 Simplified syntactic definition of the *alter*-statement

We have three alternatives here. The first two allow us to add a new attribute to an existing table and to modify the specification of an existing attribute, thus complementing the original table definition. The second one is typically used to provide some extra space for the values of the attributes concerned. The third alternative provides the means to remove some of the constraints originally imposed on the table through the create-table statement.

To demonstrate the data definition features of SQL let us consider a simplified College Database. Fig. 8.6 shows the underlying global EAR schema (a submodel of the model considered in Chapter 2).

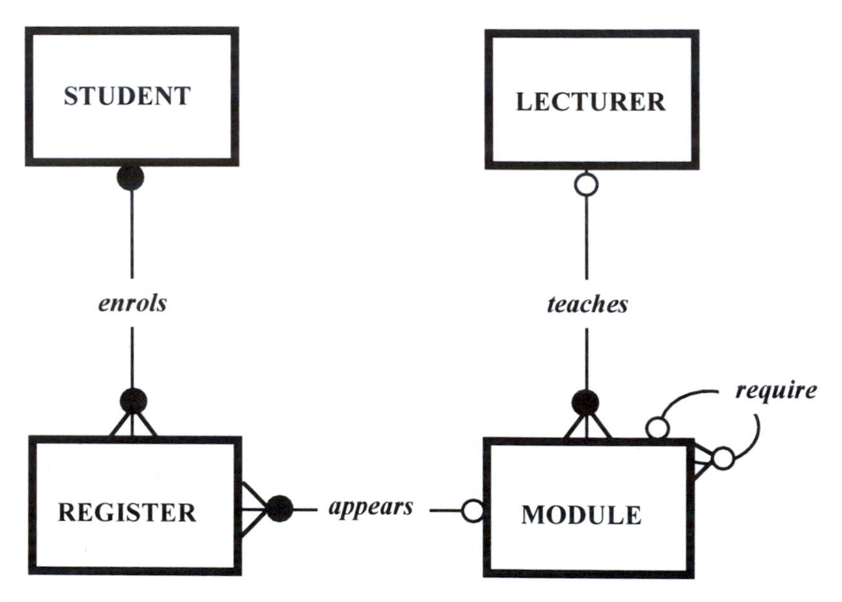

Fig. 8.6 Global EAR schema for the College Database

The schema consists of four entities STUDENT, LECTURER, MODULE and REGISTER interconnected by the following four relationships:

enrols A student may enrol for a number of modules. At any point in time (throughout the course of study) every student must be registered for at least one module and, conversely, every registration is made for a particular student.

teaches A lecturer may teach a number of modules and every module is taught by exactly one lecturer. During certain periods of time some lecturers may not be involved in teaching at all.

appears Every registration record relates to some module. While some modules may not attract any students, typically a module is attended by a number of students.

require A module may be a prerequisite for a number of other modules; some modules do not have prerequisites at all.

The relational representation of the above model consists of the four tables whose instances are depicted below, together with some description of their attributes.

STUDENT

STNO	NAME	SEX	DOB	ADDRESS	ENTRY
40986	Jones	Male	12/8/80	Oxford	1999
42331	Smith	Female	21/2/81	London	1999

STNO — 5-digit number unique to every student. **Primary key**.
NAME — 35-character string containing student names
SEX — 6-character string with the values drawn from {**Female** | **Male**}
DOB — Date of birth in the format **dd-mmm-yy**
ADDRESS — 25-character string (town/city name in this database)
ENTRY — The year in which the student has been admitted to the College, 4-character string of type date recorded in the format **yyyy**

LECTURER

ID	NAME	DOB	ADDRESS	QUAL	POSITION
55981	Adams	17/8/64	London	MEng	Lecturer
55633	Brown	21/3/56	Stoke	PhD	Reader

ID — 5-digit number unique to every lecturer. **Primary key**.
NAME — 35-character string containing lecturer names
DOB — Date of birth in the format **dd-mmm-yy**
ADDRESS — 25-character string (town/city name in this database)
QUAL — Qualification postnominal (the highest degree). 6-character string with the values drawn from {**BA** | **BSc** | **MEng** | **PhD** | ..}
POSITION — 12-character string with the values drawn from {**Lecturer** | **Senior Lecturer** | **Reader** | **Professor**}

MODULE

<u>CODE</u>	TITLE	REQUIRES	TEACHID
8049	Databases	8005	55981
8750	Compilers	8015	55633

CODE — 4-digit number unique to every course. **Primary key**.
TITLE — 9-character string containing course titles
REQUIRES — 4-digit number of the prerequisite course. A null value of this attribute denotes that no prerequisite is needed to study the course indicated by the CODE. References COURSE.CODE
TEACHID — **Foreign key** - references LECTURER.ID

REGISTER

<u>STNO</u>	<u>CODE</u>	RESULT	WHEN
40986	8049	82	15/12/00
40633	8750	75	10/12/00

STNO — **Foreign key** - references STUDENT.STNO
CODE — **Foreign key** - references COURSE.CODE
The pair (STNO, CODE) forms the primary key
RESULT — Result of the examination. Rational number from [0, 100]
WHEN — Date of examination in the format **dd-mmm-yy**

Whenever both RESULT and WHEN are null, e.g. (**42331**, **3081**, **null**, **null**) the tuple describes a particular student's attendance on the relevant course; a null value for the RESULT alone (**40986**, **3080**, **null**, **17-Jan-89**) means that the student actually attended the course but did not attempt the examination at all.

The referential integrity rules that arise from the above data design are presented in Fig. 8.7 in the diagrammatical form. The meaning of these rules is explained below.

RF 1. Every value of STNO in REGISTER must be equal to some value of STNO in STUDENT. While there may be many students registered for a particular module, each tuple must refer to an *existing* student.

RF 2. Similarly, every value of CODE in REGISTER must be equal to some value of CODE in MODULE. While there may be many modules a particular student has registered for, each tuple must refer to an *existing* module - otherwise the table REGISTER would contain the exam result obtained by a student on a fictitious module.

RF 3. Every value of TEACHID in MODULE must be equal to some value of ID in LECTURER. Many modules can possibly be taught by the same lecturer, none of them by a fictitious one.

RF 4. Every value of REQUIRES in MODULE must be equal to some value of CODE in MODULE, i.e. every prerequisite is a module itself and hence must be recorded in a tuple on its own.

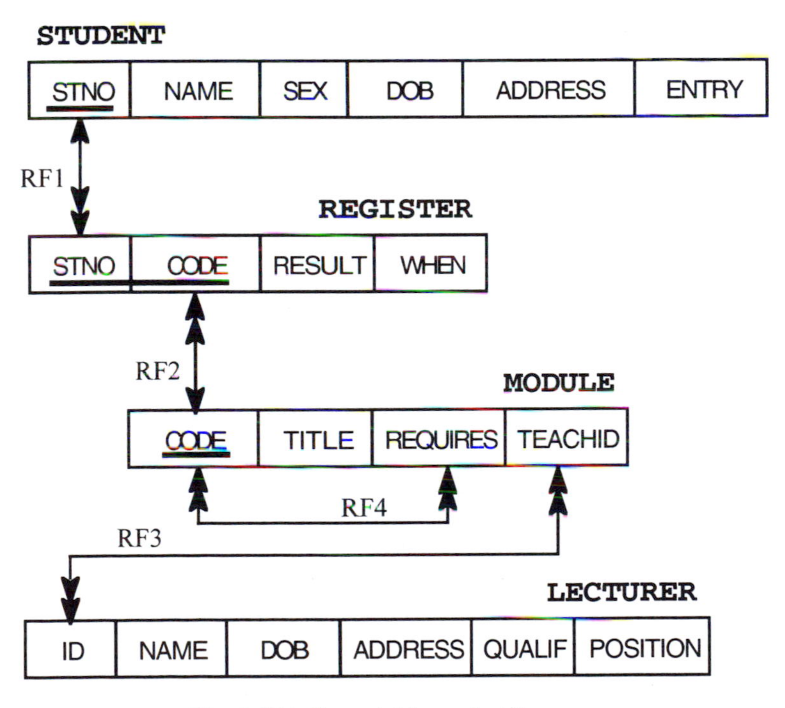

Fig. 8.7 Referential integrity diagram

The SQL code (data definition) presented in Fig. 8.8 implements the above design.

First, conventional programs almost invariably contain specifications of case-dependent, carefully designed data structures, and hence include manipulation of record addresses at (nearly) physical level to control the imposed ordering. Expressions in SQL deal, in turn, with only one simple data structure - a table. Moreover, even the known (logical) addressing system (i.e. primary keys) is not particularly important since the data can be accessed via any combination of attributes. The application programmer is thus completely insulated from physical representation of the data storage details.

Second, the algorithms that process the data structures, and in a typical conventional program play an important role, do not require here any programming effort at all. In SQL, those specifically needed are represented by single words (such as max, min, order by, etc.); in general though, the retrieving details are left to the system to execute in a way that is determined by the built-in optimizers.

Third, SQL does not support the typical program control structures, such as *if...then...else* and *while...do*, since, in principle, there is no need to employ them at all. Subject to a few simple lexical rules, the processes can be represented in an unconstrained way, flow control being left to the DBMS.

Fourth, SQL expressions genuinely represent the *set-at-a-time* approach. The objects being processed are sets (i.e. tables) rather than individual records. This, of course, contributes to the much more compact form of SQL code and with application logic more transparent than can be achieved in the block-structured programs.

8.3.1 Simple retrievals

The most fundamental form of retrieval expression combines three relational operators: *project*, *select* and *join* in the following disguise:

```
select attributes wanted
from tables that hold them
where tuples satisfy the specified conditions
```

Thus the expression:

```
select distinct CODE, TITLE from MODULE
```

is equivalent to **Π**MODULE(CODE, TITLE) whereas

```
select * from MODULE where REQUIRES= 8049
```

is equivalent to: **σ**MODULE(REQUIRES= 8049).

The meaning of the first expression is '*What are the different modules stored in the College Database*' while the second expression represents a query '*Which modules require module 8049 as a prerequisite*'. The asterisk * used in there is a convenient shorthand for '*all the attributes in the specified table*'. The phrase **select distinct** is important since SQL permits the programmer to formulate an improper project operation.

For instance:

```
select distinct QUAL from LECTURER
```

retrieves only the different postnominals from the table of the teaching staff (i.e. duplicates are necessarily discarded) while **select** QUAL **from** LECTURER extracts the whole column QUAL from the same table despite the obvious redundancy.

```
select
    [distinct]  target-list
    from        tablename [, tablename]*
    [where      predicate]
    [group by   attribute-name [,atribute-name]*
    [having     function-list]]
    order by    attribute-name [ desc ] [,atribute-name [ desc ]]*
```

Fig. 8.9 General syntax of the *select*-statement

Unlike Relational Algebra (or any programming language for that matter), SQL does not directly support relational assignment. The result of an operation cannot be assigned to a table (i.e. the operation A := *select ... from ... where ...* has not been implemented). The result of a retrieval is typically stored in a temporary buffer and, once displayed or printed, gets erased.

Sometimes the result needs to be kept though - either for reference or for further use. The simplest way to get around this problem is to create a table (or a view, depending on circumstances) by using the syntactic form given below:

```
create
    {table | view} [attribute-list]    as retrieval statement
```

Fig. 8.10 Possible implementation of relational assignment

As an example, consider the following SQL expression (note that projection is over the primary key and hence the clause **distinct** is not necessary):

```
create table BASIC_MODULE (CODE, NAME) as
  select CODE, TITLE
  from MODULE
  where REQUIRES is null
```

The retrieval statement, equivalent to the algebraic expression

$\Pi\ \sigma$MODULE (REQUIRES is null) (CODE, TITLE)

selects all the modules (or rather their CODEs and TITLEs) that require no prerequisites; then the embracing statement **create table as** puts the results into the newly created table BASIC_MODULE with attributes CODE and NAME whose specifications are copied from the original table definition. Note that the new attribute names need not be the same as the original names; if they are, there is no need to specify them after the resultant table name.

Note also the form in which the predicate in the above query is formulated. Since **null** denotes an unknown (or not applicable) value, no comparison involving a null-value is defined (i.e. any expression that evaluates to X = **null** is neither true nor false, irrespective of the value held by X). The target list that appears after **select** is generally of the form shown in Fig. 8.11.

```
target-list     ::=  attribute-list | function list
attribute-list  ::=  atribute-name [, atribute-name]*
function-list   ::=  function(attribute-name)[[   and | or ]
                     function(attribute-name)*]
function        ::=  sum | avg | min | max | count
```

Fig. 8.11 Syntax of the target-list

The use of built-in SQL functions is illustrated below by a number of examples.

Example 8.1

Find the average examination result together with the maximum examination result achieved by any student on module 3025

```
select avg(RESULT), max(RESULT)
from REGISTER
where CODE = 3025 and RESULT is not null
```

Note that all the unknown results were eliminated from the calculations.

Example 8.2
How many modules are provided by the College?

```
select count(*) from MODULE
```

The function **count**(...) simply counts the number of rows in a retrieved table, hence any attribute name can be used as an argument.

Example 8.3
How many modules require 3011 as a prerequisite ?

```
select count(*) from MODULE
where REQUIRES = 3011
```

Example 8.4
How many modules have been attended by the students?

```
select count(distinct CODE) from REGISTER
```

The predicate in the **where**-clause can be amplified (see Fig. 8.12), as the following example illustrates:

Example 8.5

```
select ID, NAME from LECTURER
    where (ADDRESS = 'Oxford' or ADDRESS = 'Abingdon')
    and DOB between '01-Mar-55' and '01-Mar-60'
    and QUAL like '%Eng%'
    and POSITION in ('Senior Lecturer', 'Reader')
```

The predicate is rather self-explanatory; the condition QUAL **like** '%**Eng**%' requesting all the values of the attribute QUAL that contain **Eng** as a substring at any position.

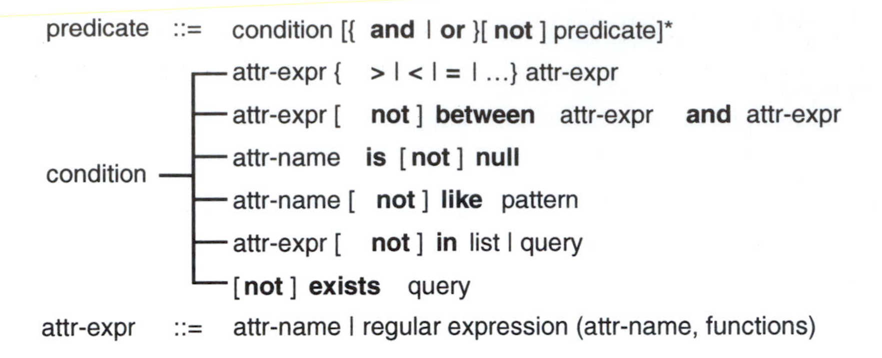

predicate ::= condition [{ and | or }[not] predicate]*
condition
attr-expr { > | < | = | ...} attr-expr
attr-expr [not] between attr-expr and attr-expr
attr-name is [not] null
attr-name [not] like pattern
attr-expr [not] in list | query
[not] exists query
attr-expr ::= attr-name | regular expression (attr-name, functions)

Fig. 8.12

SQL provides a structure for operations on subtables, i.e. groups of rows with a common value for attribute(s) indicated. The conceptual mechanism of the clause **group by**... **having**... is shown in Fig. 8.13.

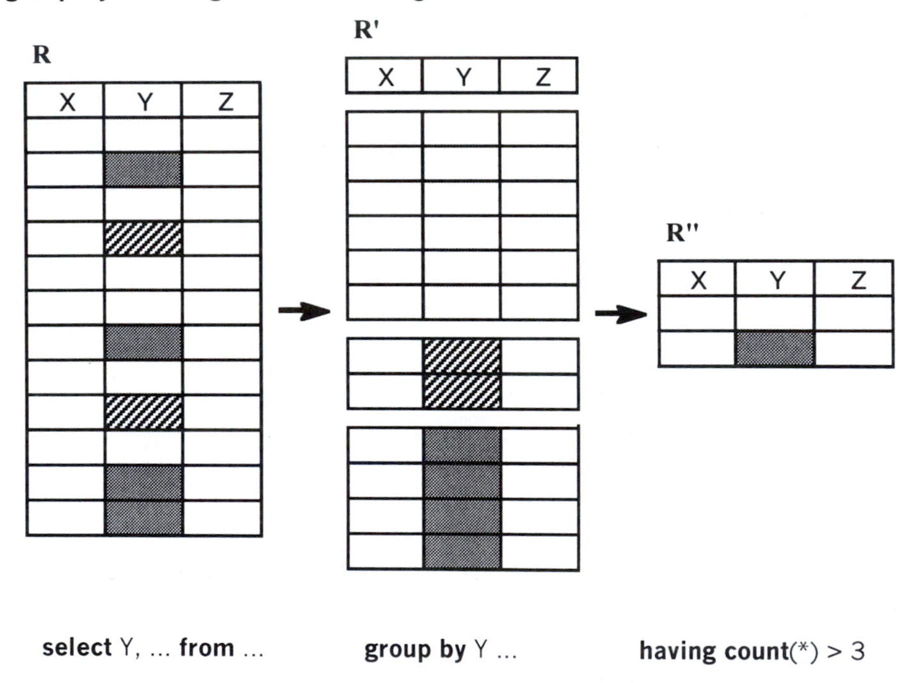

Fig. 8.13 The conceptual mechanism of **group by...having...** clause

After the initial table(s) have been obtained according to the conditions specified in the **where**-clause, the **group by** Y section partitions the result into the separate groups with their tuples having the same value for the attribute Y; finally the groups not conforming to the condition specified in the **having**-clause are eliminated.

Example 8.6
Get the lecturer identifiers of those who teach any number of modules

```
select TEACHID from MODULE
group by TEACHID
```

Example 8.7
Calculate the average examination result and find the corresponding maximum and minimum result for every module

```
select CODE, avg(RESULT), max(RESULT), min(RESULT)
from REGISTER
where RESULT is not null
group by CODE
order by CODE
```

Example 8.8
Get the average examination results for large classes

```
select CODE, avg(RESULT)
from REGISTER
group by CODE
having count (*) > 100
```

8.3.2 Retrievals from multiple tables

SQL supports two different methods of joining tables together. The first one is illustrated by the following query:

Example 8.9
Get the lecturer identifiers and names together with the titles of the modules they teach

```
select ID, NAME, TITLE
from LECTURER, MODULE
where LECTURER.ID = MODULE.TEACHID
```

The form of this join is essentially the same as that of the equivalent algebraic expression **join** LECTURER, MODULE: (ID = TEACHID). Unqualified join obviously returns a *cartesian product* of the two tables concerned.

The next example demonstrates the usefulness of synonyms when a table is joined with itself:

Example 8.10

Get the list of all pairs of students who have the same name

```
select X.STNO, X.NAME, Y.STNO, Y.NAME
from STUDENT X, STUDENT Y
where X.NAME = Y.NAME and X.STNO < Y.STNO
```

In this case two auxiliary names **X** and **Y** are used to replace the original table name (thus conceptually this is a join of **X** and **Y**); the condition required to discard all the duplicates is **X**.STNO < **Y**.STNO.

Example 8.11

Get the student numbers and names together with the results they obtained from their examinations. Present the output in alphabetical order.

```
select STUDENT.STNO, STUDENT.NAME, RESULT
from STUDENT, MODULE, REGISTER
where MODULE.CODE = REGISTER.CODE
and REGISTER.STNO = STUDENT.STNO
and RESULT is not null
order by STUDENT.NAME
```

Example 8.12

For a given student name, list all the module codes and the corresponding examination results obtained by that student

```
select CODE,  RESULT
from REGISTER
where STNO =  (select STNO
               from STUDENT
               where NAME = 'Jones')
```

The corresponding algebraic expression for this query is:

Π(**join**(**Πσ**STUDENT(NAME=**'Jones'**)(STNO),REGISTER)(STNO=STNO))
(CODE,RESULT)

The nested form allows the programmer to formulate unambiguously those kinds of query where the evaluation of the predicate depends on the result of a function.

Example 8.13

Get all those students whose results in any subject are greater than the average

```
select STNO
from REGISTER
where RESULT > (select avg(RESULT)from REGISTER)
```

In general, the nested structure

```
select ... from ... where X Θ [all | any] (select ... from ... where ...)
```

allows us to formulate a wide range of queries in a reasonably compact form. The set of comparators typically is **Θ** ::= {=, <>, >, >=, =<, <}.

The inner query may return a number of tuples, in which case the joining condition must be further qualified by use of one of the clauses **all** or **any**.

Example 8.14

Get the names of the oldest lecturers

```
select ID, NAME
from LECTURER
where DOB =< all (select DOB from LECTURER)
```

The following example introduces the concept of the **exists** operator, which in SQL is regarded as a logical function that returns the value *true* if the subquery in

```
... exists (select ... from ... where)
```

retrieves a non-empty table, otherwise the returned value is *false*.

Example 8.15

List all the students who do not take the MODULE 8011

```
select *
from STUDENT
where not exists
            (select * from REGISTER
            where REGISTER.STNO = STUDENT.STNO
            and REGISTER. CODE = 3011)
```

The above example does not, in fact, require the use of the **exists** operator; the query can just as well be expressed in the form:

SQL implementation of the 'pure' set operations *union*, *difference* and *intersection* is done through the following nested form (probably to ensure control over union-compatibility of the tables to which these operations apply):

```
select ... from ... where ...
            {union | minus | intersect}
select ... from ... where ...
```

Example 8.18

List all the modules that no students have attended

```
(select CODE from MODULE)
            minus
            (select distinct CODE from REGISTER)
```

8.3.3 Storing queries

The queries may need to be stored in some form for later (or repetitive) use. Nearly all DBMS's provide some methods of storing the SQL expressions. The one devised by ORACLE seems to be fairly representative, though by no means standard. In ORACLE the current SQL query resides in a buffer that is controlled by the database software itself; while in the buffer the query can be re-executed at any time. The buffer is emptied whenever a new query is being formulated during the interactive session (or after disconnecting from the system altogether).

The command

save *filename*.SQL

sends the content of the buffer to the user's directory and, conversely,

get *filename*.SQL

fills the buffer with the content of a given file from the directory. The files holding SQL expressions are then maintained by the host Operating System as standard textfiles and can be edited by using any of the available editors.

The communication between the DBMS and the operating system is guaranteed via

host *any permissible O/S command*

which causes temporary suspension of the DBMS and passes control to the operating system. The O/S command is executed instantly and the DBMS resumes control. Any command file (.SQL file) can be activated through

start *filename*[.SQL]

which effectively loads the buffer with the content of the file indicated by its name and runs it.

Some parts of an SQL expression (and, in principle, this applies to any string within that expression) can be substituted by a variable (*substitution variable*) whose value is undefined until the run time. Example 8.19 illustrates the point.

Example 8.19
Find the average examination result together with the maximum examination result achieved by any student on a ***given*** *module*

```
select avg(RESULT), max(RESULT) from REGISTER
where CODE = &given and RESULT is not null
```

During the interpretation phase, since the variable **&given** is undefined, the system will stop the execution and request a value to be assigned to that variable. The value supplied by the user will then replace (textually) the variable and the execution will be resumed. In this way, the query can easily be repeated many times over with a different module code for each run. In ORACLE, the substitution variable can be used effectively to replace numeral or string values, table and attribute names, file names or even the SQL restricted words. The substitution variable assumes one of the forms **&***name* or **&&***name* where *name* is any string (subject to usual naming restrictions). The ampersand sign distinguishes the *name* as the substitution variable in the SQL code.

The variable **&***name* holds the assigned value only for one (current) run, whereas **&&***name* holds its value until explicitly revoked by **undefine** *name* or by the log-off action. The command **accept** *variable-name* **prompt** 'message' facilitates a convenient way of supplying values to substitution variables, as the following example illustrates:

Example 8.20

```
rem  query1.SQL - Students who have not registered for a given module
accept CC prompt 'Enter module Code:  '
   select * from STUDENT where STNO
   not in  (select STNO from REGISTER where CODE = &CC)
accept OPTION prompt 'Which option now? :  '
start &OPTION
rem  - End of query1.SQL
```

8.4 MODIFYING THE DATA

The SQL concept of updating is based on the three elementary transactions: *delete*, *insert* and *update*. These operations change the data stored in the database but not the structures that hold the data. Each operation is indivisible and can be **commit**ted (if it terminates successfully) or **roll**ed **back**, in which case the database is brought back to the state it was in immediately prior to that operation.

Almost every implementation of SQL allows the programmer to build up a routine (a macro, a sequence or whatever the actual name might be) that encompasses a number of simple updates into a *compound transaction*. The compound transaction terminates successfully if (and only if) all of its elementary operations are successful, otherwise no change to the database occurs. This *all-or-nothing* approach provides programming means to secure the database consistency and allows the programmer to enforce referential integrity.

The issue of consistency is of paramount importance as practically all updating routines in the real systems are the compound ones. As an example, let us consider some possible updating of the College Database. Suppose a particular student occurrence is to be deleted. This would essentially require two elementary transactions to be executed as one supertransaction:

begin transaction
delete multiple tuples from REGISTER
delete single tuple from the STUDENT
end transaction

First, every tuple in REGISTER that represents the student's enrolment on a particular module is deleted; having done so (and only then) the relevant details about the student could be removed from the relation STUDENT (note that the reverse order of these deletions would temporarily violate referential integrity).

Since the relevant foreign key in the table STUDENT is defined with the clause **on delete cascade**, there is no need for that compound delete transaction as the system will automatically remove all occurrences of the foreign keys from REGISTER should the command to remove the corresponding primary key from STUDENT be activated.

The issue of a cascade deletion would thus be resolved satisfactorily, a deletion with a subsequent update would not. Suppose a particular module is no longer available and the corresponding tuple needs to be removed from MODULE. Any such a deletion would require some updating in REGISTER (rather than removal - since the students' past records cannot be changed!). A possible action may also be required with respect to the table LECTURER, e.g. in case the module occurrence being deleted is the last module taught by the corresponding lecturer - but, again, not necessarily deleting the occurrence of that Lecturer's record altogether! Since SQL does not directly support this kind of referential integrity (**on delete update**, say) a compound transaction is required.

8.4.1 Elementary updates

Simple actions on a single table, such as deletion of tuples, insertion of new tuples or amendment of attribute values are themselves considered to be atomic (i.e. indivisible) transactions. The syntactic definitions of the relevant SQL expressions are presented in Fig. 8.16; their similarity to the the retrieval statement form is, of course, intentional.

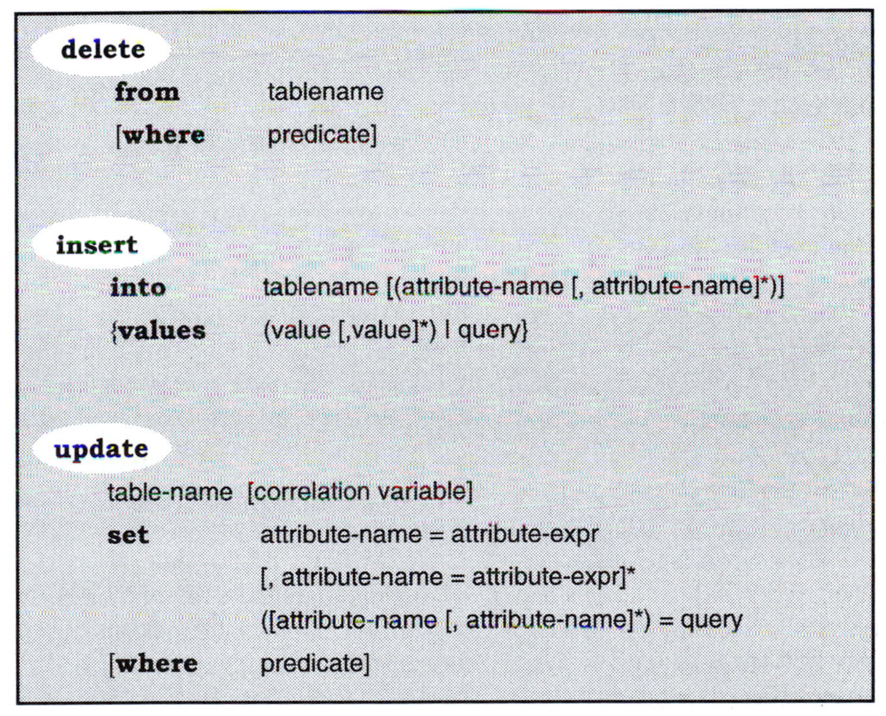

```
delete
    from        tablename
    [where      predicate]

insert
    into        tablename [(attribute-name [, attribute-name]*)]
    {values     (value [,value]*) | query}

update
    table-name  [correlation variable]
    set         attribute-name = attribute-expr
                [, attribute-name = attribute-expr]*
                ([attribute-name [, attribute-name]*) = query
    [where      predicate]
```

Fig. 8.16 Syntactic definition of *delete, insert* and *update* statements

The examples showed below demonstrate the use of the relevant forms in the context of the College Database.

Example 8.21

Delete all the exam results for the two years immediately preceding a given date

```
delete from REGISTER
where WHEN + 730 < given date
and RESULT is not null
```

Example 8.22

Delete all the students who are not registered for any module

```
delete
from STUDENT
where STNO not in
            (select distinct STNO from REGISTER)
```

Example 8.23

Add a new module on database software to MODULE

```
insert into MODULE
values ( 8027, 'DB Software', 8049, 55981)
```

Since all the attribute-values are given in the insertion statement (and in the order as they appear in the original definition) there is no need to list the attribute names after the relation name.

Example 8.24

Add a new enrolment to REGISTER

```
insert into REGISTER (STNO, CODE) values ( 40986, 3027)
```

In this case, since only two attributes are being inserted, their proper identification requires the corresponding names to be specified; the attributes whose values are not known at the time of insertion are (typically) set to **null**. Note that this simple insertion does not provide any safeguards for referential integrity.

The following example illustrates a possible use of the second form of insert. In practice, it serves as a copy facility (i.e. copy a fragment of one table into another). It can also be used as a constrained form of relational assignment since most SQL implementations would have created the target table, if it had not already existed.

Example 8.25

Copy all the entries from REGISTER *that contain excellent results in whatever subject into another table*

```
insert into WHIZZ (STNO, CODE, GRADE)
        select STNO, CODE, RESULT
        from REGISTER where RESULT > 85
```

Example 8.26

Update details of a given lecturer

```
update LECTURER
      set QUAL = 'PhD', POSITION = 'SnrLecturer'
where ID = 55981
```

More complex updating can be demonstrated by the use of a more complicated search condition as shown in the example below:

Example 8.27

Increase by 6% the salary of all those lecturers whose salary is less than the average salary within their professional group

```
update LECTURER X
      set SALARY = 1.06 * SALARY
where SALARY <
              (select avg(SALARY) from LECTURER
              where  X.POSITION = POSITION)
```

8.4.2 Advanced update actions

Let us consider the table ASSESSMENT-SCHEME whose instance is given in Fig. 8.17. The table presents details of assessment for every occurrence of MODULE in the database. Suppose we need to devise a deletion routine whose logic may be expressed as:

> *Delete a record from* ASSESSMENT-SCHEME *identified by its primary key* (MODULE, ASSESSMENT) *provided it is not the only record pertinent to the value of* MODULE (*i.e. every module has to have at least one assessment element associated with it at all times).*

By representing the relation as a set of distinct sub-sets (i.e. tuples) as illustrated in Fig. 8.17, we can easily observe that the statement:

```
delete from ASSESSMENT-SCHEME
      where MODULE = 8049 and ASSESSMENT = 'Assignment 2'
      and exists (select * from ASSESSMENT
            where MODULE = 8049 and ASSESSMENT <> 'Assignment 2')
```

will succeed, that is the record (8049, Assignment 2, SQL program) will be erased.

ASSESMENT-SCHEME

MODULE	ASSESSMENT	DESCRIPTION
8049	Assignment 1	Normlisation
8049	Assignment 2	SQL program
8049	Examination	2-hour paper
8011	Examination	3-hour paper
8012	Assignment 1	Eiffel
8012	Assignment 2	C++
8012	Assignment 3	Java
8012	Examination	2-hour paper

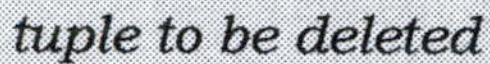

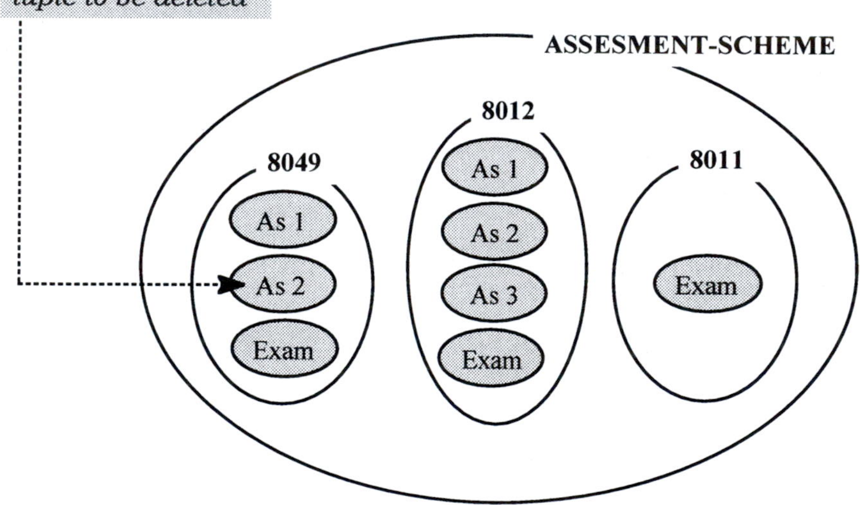

Fig. 8.17

The record (8010, Examination, 3-hour paper) cannot be deleted , hence the statement

```
delete from ASSESSMENT-SCHEME
where MODULE = 8011 and ASSESSMENT = 'Examination'
and exists (select * from ASSESSMENT-SCHEME
        where MODULE = 8011 and ASSESSMENT <> 'Examination')
```

which is essentially an identical expression save for the different variable values, fails. This example illustrates the use of the operator **exists** to verify whether a subset of a relation (identified by the query condition) is empty *after* deletion has taken place.

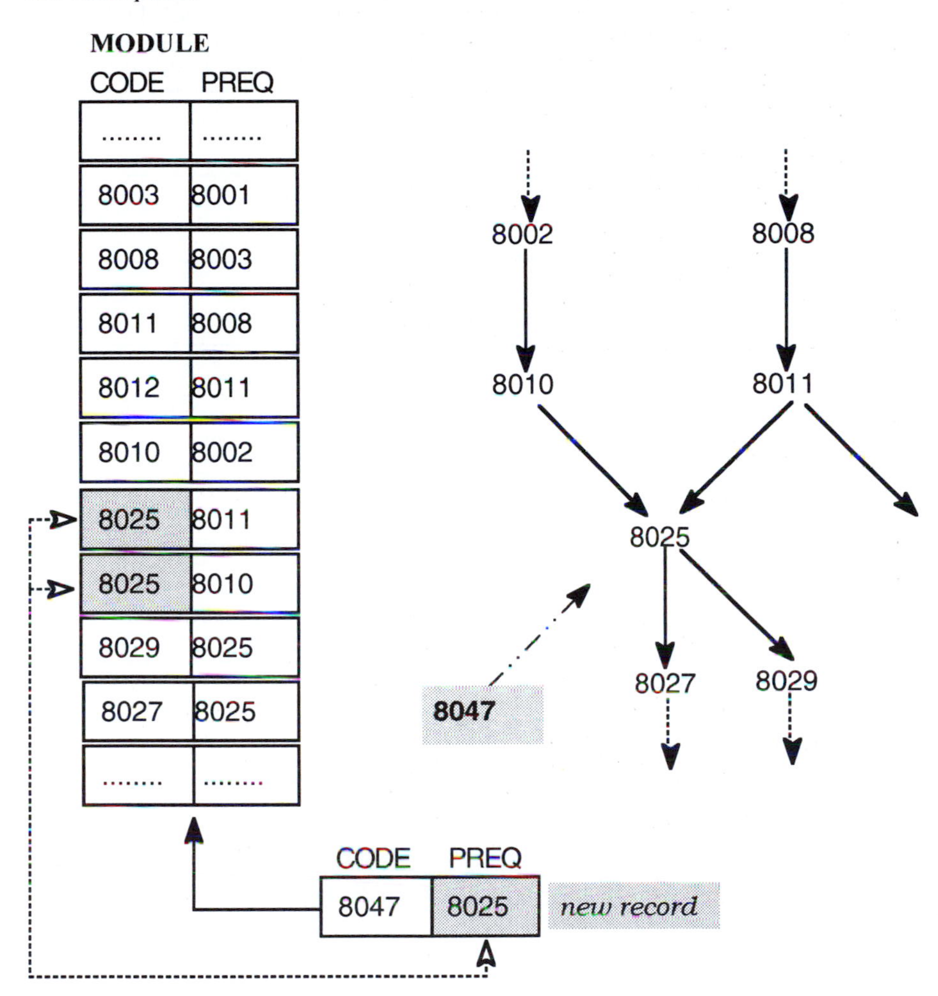

Fig. 8.18

A similar mechanism can be devised for more complex insertions. Recall the case described in Chapter 3 (Fig. 3.13 whose slightly simplified version is reproduced above as Fig. 8.18). Suppose a new module is to be inserted into the table which presents the details of the prerequisite system. If the new module 'depends' on some existing one, the latter must already be present in the table and hence the most certain way of ensuring that this is the case is actually to select its CODE-value from that table. Fig 8.19 presents the details of such design.

TRANSACTION	*Insert a record into* MODULE
pre	MODULE exists, is relationally consistent, non-empty
post	MODULE consistent with respect to the relationship *prerequisite*
input	x ← **code**$_i$ y ← **preq**$_i$
output	{success | fail} message
insertion logic	whenever a tuple <**code**$_i$, **preq**$_i$> is inserted into MODULE, the value of **preq**$_i$ must exist in CODE (i.e. the module **preq**$_i$ has its 'home' entry) - otherwise an insertion of a module **code**$_i$ with a non-existent prerequisite would have been attempted
SQL	**insert into** MODULE **select** &x, CODE **from** MODULE **where** CODE = &y

Fig. 8.19

8.4.3 Compound transactions

The original SQL facilities for transaction processing are not particularly well developed and hence the majority of transaction programming tasks are usually done either in an embedded mode (typically an SQL expression immersed in a C program) or by using some specific DBMS facilities.

As mentioned previously, a transaction is a sequence of SQL operations that the DBMS treats as a single unit of processing. The most common approach is to regard any sequence commencing with **commit** and ending with (explicit or implicit, such as disconnecting from the database) **commit** as a transaction. This, however, is not a reliable technique, particularly in a multi-processing environment. Hence the SQL 99 construct that permits the programmer to implement explicitly a given transaction logic (such as the one shown in Fig. 8.20 - *replace a module with its newer version leaving the students' results* *intact*) appears to be very attractive.

```
begin  not atomic
     begin atomic  -- replace obsolete module
        insert into MODULE
           values (8087, 'DB System Software', 8011, 55041)
        update REGISTER set CODE = 8087 where CODE=8027
        delete from MODULE where CODE=8027;
     end;
     begin atomic  -- remove inactive enrolments
         delete from REGISTER where ...
         .............................................
     end
end
```

Fig. 8.20 The concept of SQL 99 transaction

Other inventions in SQL 99, particularly those contributing to the maintenance of the integrity of databases, are the syntactic constructs allowing domain and data type definition, more extensive range of means for guarding referential integrity and the concepts of procedures, functions and triggers. The standard for SQL 99 is in its final stages of acceptance thus implementation of these facilities will undoubtedly occur soon.

8.5. EXERCISES

8.1 Use SQL to answer Exercises 4.1, 4.2 and 4.3, set in Chapter 4.

procedures need to be written with care to avoid any risk in violating integrity of the data.

More importantly, however, certain operations applicable to say FIGURES do not carry any meaning in the context of the table NETWORK - and *vice versa.* As an example, consider an operation *double_area_of* (▒). That type of operation, itself hardly expressable in terms of relational algebra, cannot be meaningfully applied to any instance of NETWORK (though its use in the context of NODES seems plausible). Furthermore, it might be possible to code *double_area_of* in a generic way, that is independently of the value for SHAPE . The operation would then be invoked whenever needed (by some other operation in the database) and the way the calculations are carried out would have been an internal matter of FIGURES and its 'private' operations.

Similarly, an operation *chain_nodes* (Li, Lj) (find a connection between two nodes on the network), not easily expressable in algebraic terms due to its recursive logic is, of course, meaningful when applied on NETWORK but senseless in the context of the other two tables.

The above example leads us to conclude that it may pay off to consider more complex (than relations) structures that would represent data objects and that these structures might have their own 'private' operations, inherently linked to the nature of those objects. It thus follows that these operations could be incorporated (in some abstracted form) into the structural definition of the objects, possibly bringing about the benefits similar to those gained from incorporating the abstractions of update operations into the structure of relations.

Certain analogy to the Object Oriented paradigm in programming is then not too difficult to accept. This triggered some research interest in attempting to transpose the principles of (and benefits from) that paradigm to databases. The key to the rationale for object oriented databases is that these principles would allow the formulation of arbitrarily complex data structures in a way that could be applied to database systems.

It has to be said, however, that Object Databases are still in their infancy. As yet, no rigorous mathematical model has been developed (or even proposed) and, consequently, no programming notation comparable to relational algebra devised. Neither has common terminology been universally accepted as the topic developed through accumulation of practice gained from sometimes very ingenious, but nevertheless individual, solutions to specific types of applications.

9.2 THE OBJECT-ORIENTED PARADIGM

The object-oriented paradigm, at least insofar as databases are concerned, relies on the notions of ***objects***, ***classes*** (object types), ***inheritance*** and ***polymorphism***. An object is considered to be an encapsulation of data (or attributes) and methods (procedures) for the manipulation of these data. Extensions of the basic object model required for persistent storage of objects are dealt with in the next section.

An object type is defined in terms of both its attributes and the operations permissible on these attributes. Object instances, or objects, are declared as being of a given type, and all objects of the same type have the same attributes and methods. It is useful, at this stage, to introduce the concept of a type-defining object (class), to which we can assign the definitions of the attributes and methods. Encapsulation, in its purest form, implies the 'privacy' of the attributes of an object. A public (i.e. system-wide) interface to the attributes must be provided by the methods specified for the type. The methods are invoked by messages to the object, and no other form of communication with the objects is permitted. On receipt of a message, an object refers to its type-defining object for the implementation of the associated method.

Inheritance allows new object types to be defined in terms of already existing types. A subtype of an existing type will inherit the attributes and methods of the supertype, but will have additional attributes and methods, which differentiate it from the supertype. Instances of a subtype may refer to its type-defining object, or, indirectly, to any of its supertype-defining objects. Figure 9.2 shows an example type hierarchy for people in a university. The instance of a postgraduate student shown has the attributes and methods defined for POSTGRADUATE, but also inherits those of STUDENT and PERSON, through the ***is-a*** links of the type-defining objects.

While a method may be defined for all object types in a branch of the ***is-a*** hierarchy, the actual implementation of the method may vary for individual types in the branch. Polymorphism allows the same message to be sent to all relevant types, but the recipient of the message can substitute a specific implementation of the method that services the message.

The next section describes the way in which these basic principles can be extended in order to capture the semantics of complex objects, and their representation in a database.

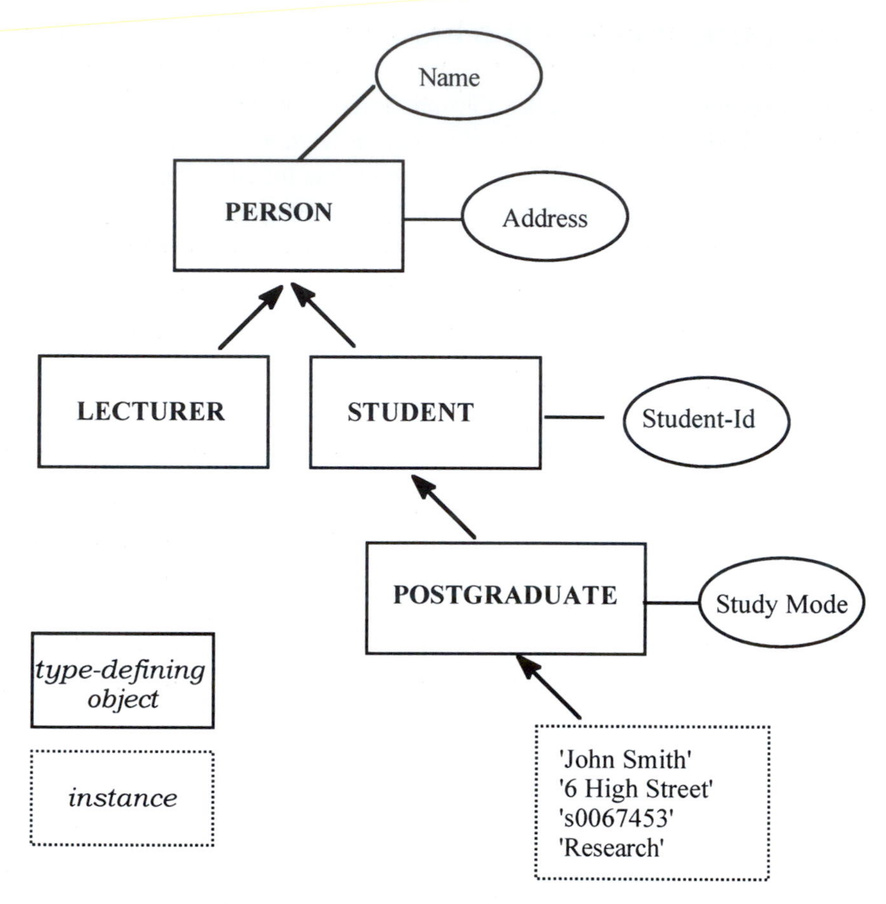

Fig. 9.2 Object-type hierarchy and instances

9.3 MODELLING COMPLEX OBJECTS FOR DATABASES

As seen in Chapter 3, the basic elements of the relational model are the attributes, drawn from suitable domains. One of the first constraints of the relational model is that the domains for attributes must be simple, or semantically non-decomposable. Typically, relational database management systems allow the specification of limited types (e.g. number, string, etc.) for the underlying domains, but more complex user-defined types seem somewhat redundant, because of the constraint on the domains. Relaxing this constraint will form the basis for the subsequent discussion. One of the first implications of this relaxation is that the operations allowed on the complex domains must be specific for each domain. In consequence, the uniform operations of the relational model will no longer be applicable and, further, the closure of the database structures under the database operations will be lost.

Definition 9.1

An object type is a subset of an expanded cartesian product of **n**, *not necessarily distinct domains* $\mathbf{D_1} \times \mathbf{D_2} \times ..., \times \mathbf{D_n}$, *such that for every element* $d^k = \langle d^k{}_1, d^k{}_2, ..., d^k{}_n \rangle \in \mathbf{R}$ *a predefined proposition* $\mathbf{p}(\langle d^k{}_1, d^k{}_2, ..., d^k{}_n \rangle)$ *is true;*
$d^k{}_i \in \mathbf{D_i}$ *for every* $i=1, 2,...,\mathbf{n}$

This is intentionally identical to Definition 3.1, for a relation. What has been altered is the nature of the domains. Each domain should be thought of as a set of values together with a set of operations that are applicable to that domain. In this sense, a domain corresponds to a data type. Domains can be arbitrarily complex, as described in the next section, and the operations applicable can be defined as necessary. An object type defined as above, becomes another domain that can be used to define further object types.

Example 9.1

Possible domain definitions that may be of interest within the university system already described are given below. At this stage the examples are described in a natural language, in order to illustrate the concepts behind complex domains. Details of the way in which object-oriented principles may be used to implement these ideas will be described in subsequent sections.

1. The domain of students is clearly of interest. This domain may be considered as a **composite** (or aggregate) of the individual domains for name, student number, term-time address, etc. Operations on the domain of students will be defined in terms of the operations already defined for these base domains. Additional operations may be defined on the new domain; e.g. change address, graduate, etc.

2. A set of students is itself a domain with normal set operations applicable. Values drawn from this domain may be used as part of other domains, such as the set of students enrolled for a given module run.

3. Postgraduate students may be defined as a domain forming a subset of the domain of students. The new domain will be a **specialization** of student, drawing additional values from other domains, having additional operations applicable, or both.

One potential problem with the formation of complex domains from simpler ones is the lack of control in the use of key attributes for identification of object instances. When domains such as sets are introduced, the problem of identity becomes even more obvious. The solution is to introduce an object identifier that is independent of the attribute values. Each instance of an object, defined on an object type is provided with an identifier (OID), which is guaranteed to be unique.

The interpretation of the principles of the object-oriented paradigm, as applied to databases, is described below.

Class (type)

> *An object type describes a domain, consisting of values and operations, valid for the type. The operations valid for a given type are called methods.*

Object instance

> *Instances of an object type must take values from the domain specified for the type, and have access to the methods of the type. Instances may be referred to as objects, and are provided with a unique identifier, called the Object Identifier (OID).*

Specialization

> *Object types form an **is-a** hierarchy of specialization and generalization. Base types will include **integer** , **string** , etc. Types, such as OBJECT and SET are included to allow the specification of 'roots' for the hierarchy. Specialization is the process whereby a component domain of a new object type is defined to be a subset of an existing domain. The new object type is known as a subtype of the original. Other component domains of the subtype will distinguish the subtype from its supertype. The subtype inherits the domain values and methods of the supertype.*

Aggregation

The domain of an object type may be formed by aggregating domains of other object types. This means that the values and operations of the component domains are available to the new object type, as well as operations defined on the new domain. Aggregation is the process whereby a set of existing domains are combined to form a new domain. An instance of the type defined on this new domain comprises a set of values drawn from the component domains.

Association

Object types may be associated with other object types, using the process of association. This corresponds with the usual relationship types, already discussed for relational databases. An instance of a relationship will be represented by the inclusion, in at least one of the associated objects, of a domain of OIDs of the other associated objects.

Invocation of Methods

The state of an object instance is defined by the values drawn from its component domains. Changes in these values are achieved through the invocation of methods specified either for the component domains, or the domain of the object type. The receipt of a message invokes methods specified for the instance.

Some of these concepts are illustrated in Fig. 9.3 where possible instances of object types LECTURER, MODULE and MODULE_RUN are shown. MODULE_RUN has attributes for the date of the run, and is associated with a LECTURER, a MODULE and a set of STUDENT. These associations are recorded as the OIDs of an instance of LECTURER, an instance of MODULE and an instance of a set of STUDENT respectively (the set of STUDENT will be described in a later section).

Each object instance in the figure also contains a pointer to its type-defining object, which is used to specify the domain of the type, including its methods.

The difference between aggregation and association is possibly unclear. In the above example, it is possible to form a 'pure' aggregation of domains for the type MODULE_RUN by including the actual domains for LECTURER, set of STUDENT and MODULE.

However, this means that it would no longer be possible to refer to a LECTURER without accessing the MODULE_RUN object. In practice, it is better to represent the aggregation by using OIDs of the aggregated types, even though this representation is identical to that for association.

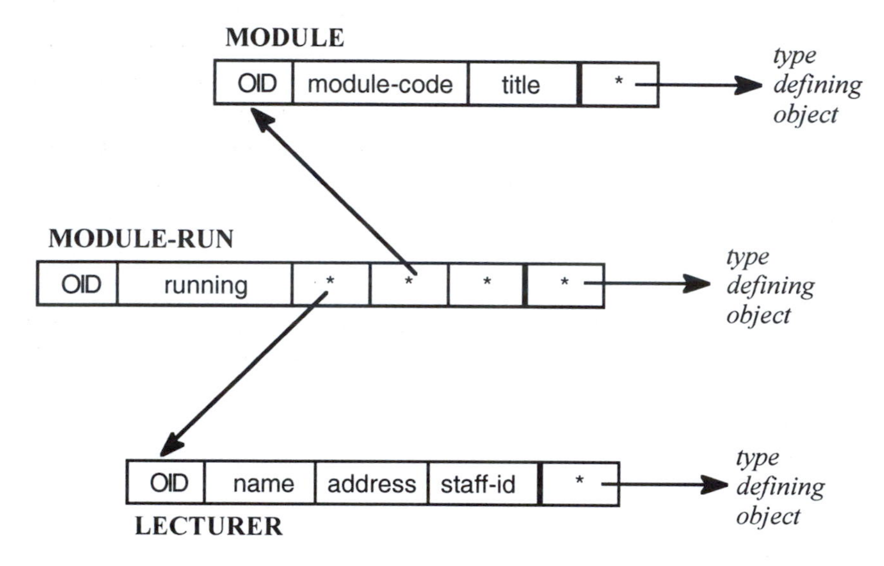

Fig. 9.3 The structure of object instances

Data design for object-oriented databases follows the principles of EAR modelling, but must be extended to allow the representation of the concepts illustrated above. An association between two object types will be modelled in the same way as a relationship between two entity types.

Object models typically do not show the methods associated with the object types, or the way in which messages may be communicated. Behavioural models may be used in conjunction with the data model to represent such information. Such modelling is not within the scope of this chapter.

9.4 DATA DEFINITION AND MANIPULATION

This section describes some of the implementation issues associated with 'pure' object-oriented databases. Although there is no universally accepted standard for object technology, the principles illustrated here are sufficient for this chapter.

Data definition is based on the specification of object types. Object types will specify the values and methods of the underlying domains, and will be built upon predefined, base types. The aim of this section is to show the principles of data definition and manipulation in an object database. The notation used is very loosely based on that of OPAL. Simple domains such as number and string are assumed, but two base types need some initial explanation. OBJECT is predefined as the root of the object type hierarchy and SET is predefined to encapsulate the methods of set manipulation. SET is not the only data structure which may be used; LIST, BAG, and others may also be included, but are not necessary for this illustration.

Application-specific object types can be defined using specialization, association and aggregation of other types. The following examples show how this can be achieved:

```
OBJECT subtype ADDRESS [
    [street#, NUMBER],
    [desc, STRING],
    [postcode, STRING]].
```

An object type ADDRESS is defined as the aggregation of three base types. Note that 'street#' is not an object type, and can only be referenced through methods associated with ADDRESS.

```
OBJECT subtype PERSON [
    [name, STRING],
    [contact, ADDRESS]].
```

PERSON is a further aggregation of a base type and the user-defined type ADDRESS. Note that 'contact' holds the OID of an address, since an instance of the type ADDRESS is an object in its own right. Aggregation in this case is implemented in the same way as an association.

```
PERSON subtype STUDENT [
    [student#, STRING]].
SET subtype SSET
    member STUDENT.
```

Local variables needed for the method are declared as |*local_variable*|. PUT and GET methods are predefined for each attribute of an object type. PUT stores a value in the attribute slot, and GET retrieves the value from the slot. ADD is a predefined method on a set type that adds a new member to the set. The SELF allows reference to the set instance that receives the message. Note that OPAL does not name its methods, instead the type name and the external parameter list identify them. External parameter labels are tied to internal method labels at the start of the method. Messages consist of the object instance name and parameter list:

```
OID_OF_SET_OF_ALL_STUDS ADD_STUD#        : 'e009'
                        ADD_SNAME        : 'Lee'
                        ADD_SCONTACT     : '6, High Street'
```

Method 2: Enrolling a student for a module run

A method is required to take an existing STUDENT object, and add it to an existing MODULE_RUN object. For this illustration, it can be assumed that the student# of the student is 's0064537'; the module run is identified by a module# 'm08005' and a running date 01/09/00. The set type MRSET for module runs has been defined, together with an instance of this type, OID_OF_ALL_MOD_RUNS.

It is worth considering the message associated with the method as a way of identifying the requirements for the implementation. In this case, the message would be sent to the set of all module runs, since at this stage the OID of the particular run of interest is unknown. The message would therefore be:

```
OID_OF_ALL_MOD_RUNS    ADD_STUD : 's0064537'
                       MOD# : 'm08005'
                       RUN_DATE : '01/09/00'
```

This raises a general point regarding the apparent lack of primary key attributes in an object-oriented database. It is clear from this example that the user of this method requires the primary key attribute of MODULE. It would be pointless to use this method if 'm08005' did not uniquely identify a module, or if the combination of module# and date did not uniquely identify a module run. While the OID of an object instance uniquely identifies that instance for the DBMS, it cannot be used by the users of the database, who will still rely on attribute values to identify instances. The method will have to find the OID of the module, identify a particular module run from this OID and the date, and add the OID of the student to the set 'enrolled' for that module run.

Method definition

```
METHOD : MRSET
    ADD_STUD : STUD_PARAM
    MOD# : MOD_PARAM
    RUN_DATE : DATE_PARAM

    |MOD_OID, MOD_RUN_OID, STUD_OID, ENROLLED_OID|

    MOD_OID := OID_OF_SET_OF_ALL_MODS
            DETECT :    [ :MX       | MOD_PARAM = MX GET_MODULE#].
    MOD_RUN_OID := OID_OF_ALL_MOD_RUNS
            DETECT:     [ :MRX    | MOD_PARAM = MRX GET_MODULE &
                                  | DATE_PARAM = MRX GET_RUNNING].
    STUD_OID := OID_OF_ALL_STUDS
            DETECT :    [ :SX     | STUD_PARAM = SX GET_STUDENT#].
    ENROLLED_OID := MOD_RUN_OID GET_ENROLLED
    ENROLLED_OID ADD STUD_OID.
```

DETECT is a predefined method for any SET class, where MX, MRX and SX are range variables over the specified set. The result of the DETECT is to return the OID of the first object instance for which the specified condition is true. The first DETECT returns the OID of the module instance with module# = 'm08005'; the second returns the OID of the module run instance with running date = '01/09/00' and the module reference = the OID of the module 'm08005'; and the third returns the OID of the student instance with student# = 's0064537'.

These examples illustrate the way in which data can be manipulated in object-oriented databases. Data retrieval depends on the existence of primary key attributes, and is performed by the sequential processing of files of records, accessed by the global set variables for the various object types in the database.

9.5 OBJECT-RELATIONAL DATABASES

This section describes approaches to integrate the benefits of object oriented database technology with the strengths of the relational model. Such hybrids are based on the relation as the main structure for data, but include additional features to allow the inclusion of complex data structures.

It has been seen that object-oriented databases rely on the ability to reference object structures through pointers, and on the use of inheritance to allow the reuse of data structures. Both of these facilities are based on the provision of facilities to define types, which is not inherent to the relational model. Relations in the relational model do not correspond with object types in an object-oriented database. A better analogy is that the relations correspond with the sets of object instances seen in the previous section. In order to define a type hierarchy, and to specify methods applicable to all instances of a type, it is necessary to add type definitions to the relational database model.

The following examples of an object-relational database are based on Oracle 8. The examples are only used to illustrate the problems and potential solutions associated with combining the relational database model with the principles of object orientation. The examples will use the same student database structures as were used in the previous section.

```
create [or replace] type address_t as object (
    street# varchar2(3),
    desc varchar2(40),
    postcode varchar2(7));
```

The creation of a type is independent from the creation of a table, or relation. It is possible to create 'column types', which can be used as nested relations. The above type will be used in this way, i.e. no table as ADDRESS will be created.

```
create type person_t as object (
    name varchar2(50),
    contact address);
```

The above definition corresponds to the concept of aggregation. Without a table for addresses the address cannot be considered as an object in its own right, but the nested relation provides a different representation than the association of the object-oriented model.

Inheritance is implemented through a similar mechanism, where the type definition is used as a column type. However, as will be shown later, the type definition may include methods, which can be accessed by the inheriting type.

```
create type student_t as object (
    student_id varchar2(8),
    person person_t);
```

'student_t' is defined as a type which includes person_t as an inherited structure.

This is not equivalent to the natural inheritance of the object-oriented model, but provides a means of reusing predefined types.

It is useful at this stage to define a table, based on the student_t type specified above. This will clarify some of the issues described so far.

```
create table student of  student_t(
    student_id  primary key);
```

It is not necessary to define a table of type person_t, since all instances will be of subtypes (only student_t, so far). Note that it is still possible to define primary keys in the tables, and these will be enforced by the relational DBMS. Data can be added to the student table, allowing for the embedded structures, as shown below.

```
insert into student values
        ('s0063457',
         person_t('Smith',
                 address_t('73', 'High Street', 'PZ4 9GH' ));
```

Object types for MODULE and MODULE_RUN are defined below. For this example, the lecturer for a module run has been ignored.

```
create type module_t as object (
    module_id varchar2(8),
    title varchar2(30));
create type module_run_t as object (
    running date,
    module ref module_t);
```

Set types are not included in Oracle 8, which means that the set of students registered for a module run needs to be represented as a many-to-many relationship. It is necessary to define a type, enrolled_t, which will be used to store the instances of students enrolled for module runs.

```
create type enrolled_t as object (
    student ref student_t,
    module_run ref module_run_t);
```

Instances of object types are given an OID by the system. Thus each row of the tables declared on the above types is uniquely identifiable by this OID. The 'ref' construct is used to specify that the value contained in an attribute is to be the OID of an instance of the associated type. Thus, a table defined on the type enrolled_t will have two columns; the first will hold the OID of an instance of student_t, and the second, an OID of an instance of module_run_t. Table definitions for the above types are given below.

```
create table module of module_t;
create table module_run of module_run_t;
create table enrolled of enrolled_t;
```

Adding data to tables which include 'ref's clearly depends on the existence of the instances being referred to. In order for a student to enrol on a module run, the student, the module and a particular module run must exist in the database.

```
insert into module values ('m08005', 'Warehouse Management');
insert into module_run
    select '01/09/00', ref(m)
    from module m
    where m.module_id = 'm08005';
```

The OID for the module with module_id 'm08005' is obtained from the module table, using the primary key. A new row in the module_run table now contains the running date and the OID of this module. This row will also have an OID, which can be used to add data to the enrolled table.

The example below shows the addition of a row representing the student with student_id of 's0063457', enrolling for the module run given above.

```
insert into enrolled
    select ref(s), ref(r)
    from student s, module m, module_run r
    where s.student_id = 's0063457' and
          m.module_id = 'm08005'and
          r.running = '01/09/00' and
          r.module = ref(m);
```

While the OIDs of the object instances are used internally, the need for primary keys is obvious from the above example. MEMBER methods can be defined as part of object type definitions. The body of the method is defined as a function, which can then be called as if the function name were another attribute label. The example below illustrates this concept.

```
create type module_run_t as object (
    running date,
    module ref module_t,
    member function duration (start_date in date) return number,...);
```

```
create type body module_run_t as
    member function duration(start_date date) return number is
    begin
        return(sysdate – start_date);
    end;
```

The table module_run would be defined as before. Note that each row of the new table now has an 'extra attribute', namely 'duration', which will return a value as defined by the method: **select** r.duration(r.running) **from** module_run r;

We have briefly described in this Chapter the main concepts underlying object databases, and looked at some of the implementation issues of these. The benefits expected from the object-oriented approach could be identified as efficient means to store and manipulate complex objects, particularly when the nature of the application system being developed is not quite compatible with relational representation and /or algebraic processing. It should be emphasized, however, that in a large number of applications the issue of complex objects is not relevant, and that the object databases have not yet reached the level of maturity (or indeed acceptance) comparable to that of the relational approach .

CHAPTER 10

SQL extensions

10.1 INTRODUCTION

We have already mentioned in Chapter 8 that SQL was originally perceived as a database language rather than a programming language and therefore its standard version did not support certain program (nor data) structures that were meant to be handled by the host programming language. In consequence, pure SQL does not provide any means (of course) to program operations that do not conform to the principle of relational closure and, perhaps more importantly, the means to program complex update transactions.

The lack of support for the latter is particularly disconcerting, especially in situations where referential integrity may not have been addressed at the data definition level, or in the case of enforcing particular integrity constraints, i.e. constraints imposed by the nature of the application being developed. Individual software vendors attempted to address these issues by adding non-standard extensions to their implementations of SQL. These extensions typically cover a range of database processing functionality - from simple output formatting, interfaces to software packages, programming language pre-processors through to incorporating full operational facilities of algorithmic programming languages.

The purpose of this chapter is to present an overview of such an extension using as an example PL/SQL - the system developed by ORACLE to support, amongst others: user-defined data types, conditional and looping structures, exception handling, functions, procedures, triggers and packages. The next two sections are based on Oracle*PL/SQL Version 2. Their aim is purely illustrative for the notions, concepts, techniques, etc. are only applicable to one particular DBMS and not necessarily shared across the whole range of relational products. The readers interested in the technical details are referred to *PL/SQL Users Guide and Reference* (Oracle Corporation 1992).

10.2 BASIC PROGRAMMING STRUCTURES

PL/SQL can be used in just the same way as 'pure' SQL (i.e. interactively) but its more natural use is to code the components of command files (or inbuilt fragments of packages) and get them compiled and stored as database objects.

The smallest meaningful unit of code is the PL/SQL block whose structure is shown in Fig. 10.1. (the forward slash activates the execution of the block).

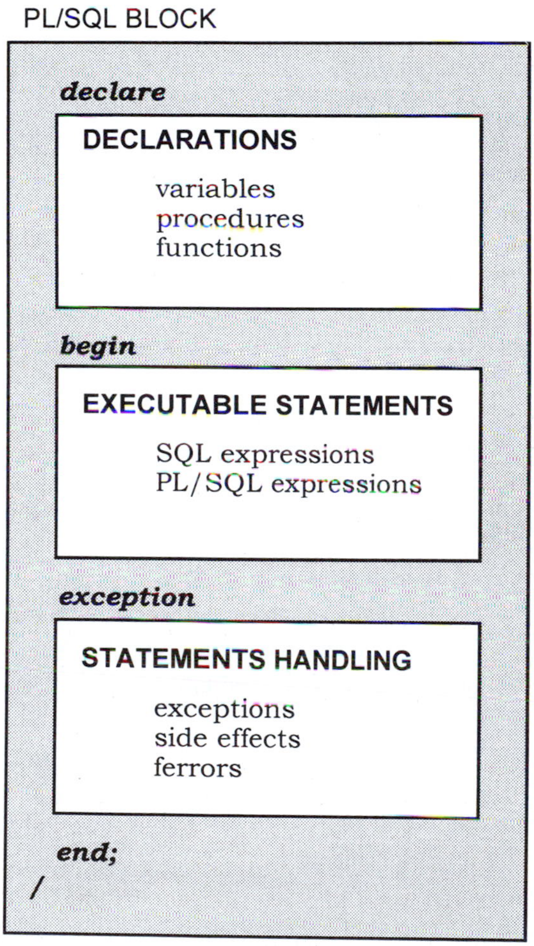

Fig. 10.1

A PL/SQL block can be thus written into .sql file which can, of course, contain other SQL expressions as well as other PL/SQL blocks; block nesting is allowed.

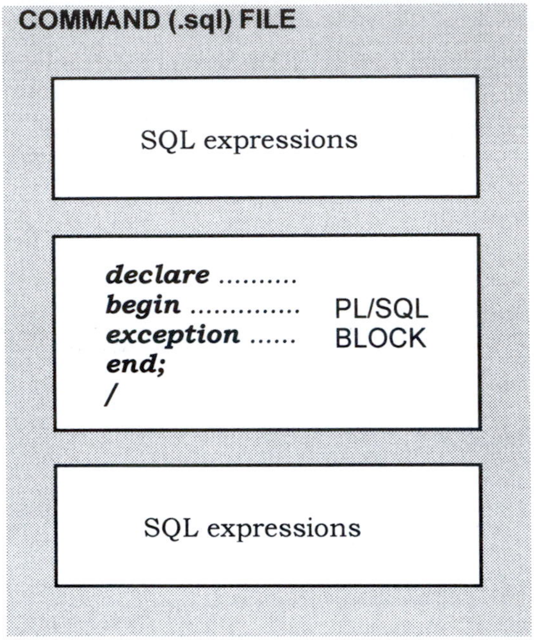

Fig. 10.2

The communication between PL/SQL block and pure SQL is implemented in two ways: through variables declared within the block or auxiliary tables (i.e. tables that do not form part of the underlying database design), which can then be processed by SQL. Declaration of variables requires basic pre-defined data types; these include types such as integer, real, char, date, boolean, and user-defined types, records (including nested records) and tables.

Fig. 10.3 illustrates the way simple variables can be declared; the variable **x** is declared explicitly but the other two (**id** and **who**) inherit their types from the attribute type definition in the table STUDENT thus ensuring type compatibility in the corresponding select statements.

PL/SQL introduces the concept of a record variable as illustrated in Fig. 10.4. The record variable lec-rec inherits the structure and the component data types from the relevant table definition thus lec-rec.Id correspond to LECTURER.ID and so forth.

```
declare
    x number;
    id STUDENT.STNO % TYPE;
    who STUDENT.NAME % TYPE;
    ........
begin
    select STNO, STID into id, who
    from STUDENT where ...........;
    ........
end:
/
```

Fig. 10.3 An example of variable declaration

```
declare
    ........
    lec-rec LECTURER % ROWTYPE
    ........
begin
    select ID, NAME DOB, ADDRESS, QUAL, POSITION
    into lec-rec
    from LECTURER  where .........;
    ........
end;
/
```

Fig. 10.4 Simple record declaration

User-defined record (possibly nested) can also be declared. This is achieved through type specification which is then employed to define records of various types and structures. Fig. 10.5 shows the declaration of a composite record and illustrates the way the record and its component parts are referenced.

Another composite data type introduced in PL/SQL is an unbound one-dimensional array (the actual name in PL/SQL is a table), i.e. a structure made of an unnamed column uniquely indexed by a pseudo-column of type binary integer.

Fig. 10.6 shows two types **markTableType** and **moduleTableType** being specified and subsequently used for declaration of two PL/SQL tables MODULE-TAB.

```
declare
type timeT is record
      (minute integer,
      hour integer);
type location is record
       (building char(2),
       room integer);
type tutorialT is record
      (day date,
      time timeT,
      venue location,
      subject MODULE.TITLE% TYPE);
seminar tutorialT;
.....
begin
      seminar. day := '15-Feb-2000';
      seminar.time.minute := 15;
      seminar.time.hour := 9;
      seminar.location.building := 'Newton';
      seminar.location.room := 101
      select TITLE into subject from MODULE where ...
      .....
end;
```

Fig. 10.5 Composite record declaration

and MARK-TAB . The counter **i** is declared as binary integer and initialled to 0; it serves as an index referencing the entries in both tables.

This example demonstrates the use of a loop to facilitate copying the values of STNO and RESULT from the database table REGISTER into STUDENT-TAB(i) and MARK-TAB(i), respectively. The loop is controlled by the cursor **regcurrent**.

```
declare
type markTableType is table of REGISTER.RESULT% type
index by binary_integer;

type studentTableType is table of REGISTER.STNO% type
index by binary_integer;

STUDENT-TAB studentTableType;
MARK-TAB markTableType;
i  binary_integer := 0;
...........

for regcurrent in
            (select STNO, RESULT from REGISTER
            where CODE = 8049)
    loop
    i := i +1;
    STUDENT-TAB (i) := REGISTER.STNO;
    MARK-TAB (i) := REGISTER.RESULT;
    end loop;
...........
```

Fig. 10.6 Composite record declaration

```
......
select RESULT, STNO into X, id from REGISTER
where CODE = 8049 and STNO = 40633;
......
if X >= 40
    insert into MESSAGE values ('Module 8049 passed', id);
elsif X >= 20
    insert into MESSAGE
     values ('Re-examination granted on 8049 ', id);
else
    insert into MESSAGE values ('Module 8049 failed', id);
end if;
......
```

Fig. 10.7 An example of a condition statement

Conditional statement has been implemented in PL/SQL in a typical way. The fragment of code depicted in Fig. 10.7 converts the numerical examination result of a given student on a given module onto a three-point scale (pass, re-sit, fail) and stores the outcome in an auxiliary table MESSAGE (PROMPT: varchar2(25), STNO: number(5)). Incorporating conventional programming structures into the SQL environment opened up possibilities to program queries that might not be expressable in pure SQL (since, for instance, they would not conform to the principle of relational closure or attempt to impose an order on a relation) yet they are meaningful in the context of a given database. Some of such queries can be formulated through the use of cursors.

Consider, for example, a query that retrieves five highest scorers on a module. Clearly, the query is not within the scope of relational algebra (i.e. no functional composition of any combination of algebraic operations is equivalent to its meaning) and its representation in pure SQL would have been extremely awkward leading to abnormal waste of memory (e.g. subsequent retrievals of the highest scorer from the copy of the relevant relation updated by the previous retrieval by deleting the previous highest scorer).

```
declare
  cursor C is
       select STNO, RESULT
       from REGISTER where CODE = 8049
       order by RESULT desc;
  ID  integer;
  MARK  real;

  begin
    open C;
    for i in 1.. 5 loop
       fetch C into ID, MARK;
       exit when C%NOTFOUND;        -- if < 5 students enrolled on 8049
       insert into WHIZZ values ( ID, MARK);
    end loop;
    close C;
  end;
/
select * from WHIZZ order by GRADE desc;
```

Fig. 10.8 Using a cursor in a loop

However, the query can be expressed quite naturally in PL/SQL as demonstrated in Fig. 10.8. The mechanism of retrieval is based on the cursor C whose declaration makes it possible to reduce the relation REGISTER to the relevant tuples, and to subject it to the logical ordering with regard to the value of RESULT. The operation **fetch** copies the values of STNO and RESULT from the tuple currently pointed to by the cursor into the program variables ID and MARK which, in turn, are inserted into a predefined table WHIZZ. The loop may be terminated prematurely (and without any side effects) in case there is less than five students registered for the module.

PL/SQL has also the programming means to deal with run-time errors, warnings and side-effects, generically termed exceptions. Some exceptions (such as, INVALID_NUMBER, INVALID_CURSOR, NO_DATA_FOUND, ZERO_DIVIDE, VALUE_ERROR, etc.) are defined by the system but many others can be defined by the programmer. The user-defined exceptions must be declared within the PL/SQL block to which they apply; all of them require exception handlers (see Fig. 10.9). Exceptions are raised whenever an error occurs; the system defined automatically while those user-defined must be raised explicitly.

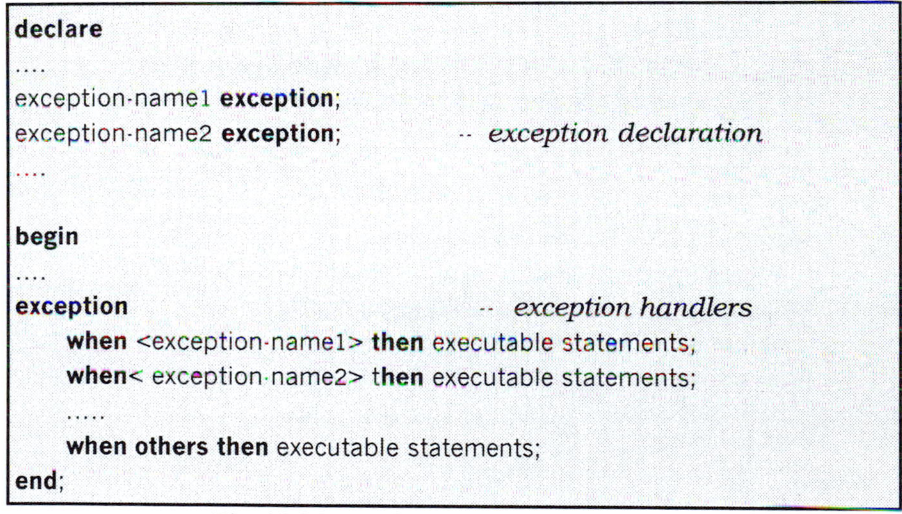

```
declare
....
exception-name1 exception;
exception-name2 exception;                  -- exception declaration
....

begin
....
exception                                   -- exception handlers
    when <exception-name1> then executable statements;
    when< exception-name2> then executable statements;
    .....
    when others then executable statements;
end;
```

Fig. 10.9 Exception declaration and handlers

Fig. 10.10 presents a fragment of code (an example due to G Swindon, 1993) that deals with an insertion of an additional element of assessment into the table ASSESSMENT-SCHEME (MODULE, ASSESSMENT, DESCRIPTION) (see

```
................
accept X  prompt 'MODULE CODE: '
accept Y  prompt 'ASSESSMENT TITLE: '
accept Z  prompt 'DESCRIPTION: '

declare
       ERR_MESSAGE       char(100);
       ASGN              char(15);
       MCODE             integer;
       DUPL_ASSGN        exception;
       INVALID_CODE      exception;
begin
      select CODE into MCODE from MODULE where CODE = '&&X;
      if MCODE = '&&X' then
         begin
           select ASSESSMENT into ASGN from ASSESSMENT-SCHEME
           where MODULE = '&&X' and DESCRIPTION = '&&Z' ;
           if ASGN = '&&Y' then raise DUPL_ASSGN end if;
           exception
                when NO_DATA_FOUND then insert into RUNINFO
                values (sysdate, 'Invalid module code');
           end;
       end if;
exception
      when NO_DATA_FOUND then insert into RUNINFO
         values (sysdate, 'Invalid module code');
      when DUPL_ASSGN then insert into  RUNINFO
         values (sysdate, 'Assignment already exists');
      when others then
         ERR_MESSAGE = substr(SQLERRM,1,100);
         insert into RunInfo values (sysdate, ERR_MESSAGE);
commit;
end;
/
................
```

Fig. 10.10 An example of using exception

Fig. 8.17 on page 172) which is connected to the table MODULE (CODE,

TITLE, REQUIRES, TEACHID) (see page 152) via the referential integrity constraint ASSESSMENT-SCHEME. MODULE *references* MODULE.CODE.

The program makes use of two system variables: **sysdate** (stores today's date) and SQLERRM (to which the system prompts are directed automatically) as well as an auxiliary two-column table RUNINFO which holds the program-generated messages about the results of processing.

10.3 PROCEDURES, FUNCTIONS AND TRIGGERS

Procedures and functions are objects that consist of set of PL/SQL expressions grouped together and executed as a unit. A simple example of a PL/SQL procedure, composed of one block (note the absence of the keyword declare), is presented in Fig. 10.11.

```
procedure upgrade (student_id, integer, module_id integer, amount real)
is
        old_mark real;
        new_mark real;
        out_of_range exception;
begin
            select RESULT into old_mark from REGISTER
            where STNO = student_id and CODE = module_id;
            new_mark := old_mark + amount;
            if new_mark > 99
                        then raise out_of_range
                        else
                        update ACCOUNT set RESULT = new_mark
                        where STNO = student_id and CODE = module_id;
            end if;
exception
             when out_of_range then
             ..............................
end upgrade;
```

Fig. 10.11 An example of a procedure

A procedure can be called from anywhere within the block in which it was declared. A possible call for the above example might be:

```
upgrade (who, on_what, by_how_much)
```

Naturally, the types of the formal and actual parameters must necessarily agree; the latter must also have their values suitably instantiated prior to the procedure call. While procedures perform some prescribed actions, functions return values and are typically called as part of an expression.

A special type of a PL/SQL procedure is a database trigger, i.e a procedure associated with a database table that runs automatically when a triggering event (typically insert, update or delete) occurs. There are two types of triggers whose mode of operation depend on whether some action is supposed to take place before or after the triggering event. **Before**-triggers execute their action prior to the triggering event since their primary function is to determine whether that event should complete at all. In contrast, **after**-triggers execute the pre-defined action immediately after the triggering event terminates.

```
create trigger <trigger name>
        before insert or update of <specification of table attributes>
        for each row
        when <boolean expression>   -- conditions activating the trigger
declare
        ..........                  -- types, variables, exceptions
        ..........
begin
        ..........                  -- PL/SQL expression
        ..........
exception
        ..........                  -- handlers
        ..........
end;
```

Fig. 10.12 Database trigger schema

An interesting development in PL/SQL is the concept of a package, i.e. a database object that groups types, objects, procedures and functions. It is typically composed of two parts: the specification of types, objects, procedures that are accessible externally and the package body, which is the actual implementation of the first part and cannot be seen by any application referring to the package. Packages offer certain advantages such as those pertaining to the modular approach to application design, information hiding, code re-usability, transaction processing and, in the context of Oracle DBMS, better performance.

CHAPTER 11

Case study - implementation

11.1 INTRODUCTION

The aim of this Chapter is to illustrate some of the theoretical concepts discussed earlier by developing a simple application based on a small subset of the University Database. The application is meant to be accessible from a network browser and will be constructed by using publicly available software freely obtainable for non-commercial use.

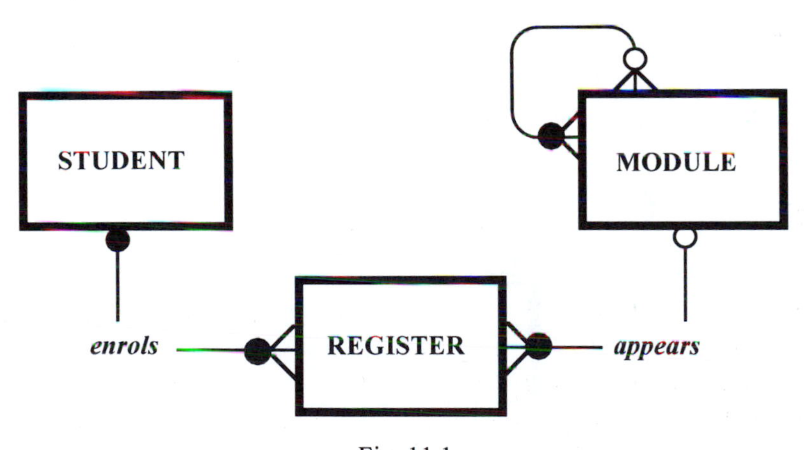

Fig. 11.1

The underlying logical data model for the system being developed is shown in Fig.11.1. We have slightly changed the model presented in Chapter 2 to incorporate the requirement that any module may require many pre-requisites as well as be a pre-requisite for a number of other modules; hence the *m:n* intra-relationship imposed on the entity MODULE.

The table representing the entity MODULE needs to be normalized to BCNF (see Fig.11.2); this effectively creates a separate table PREQ which holds the details of the module pre-requisite hierarchy, i.e. represents the graph of connections between the modules.

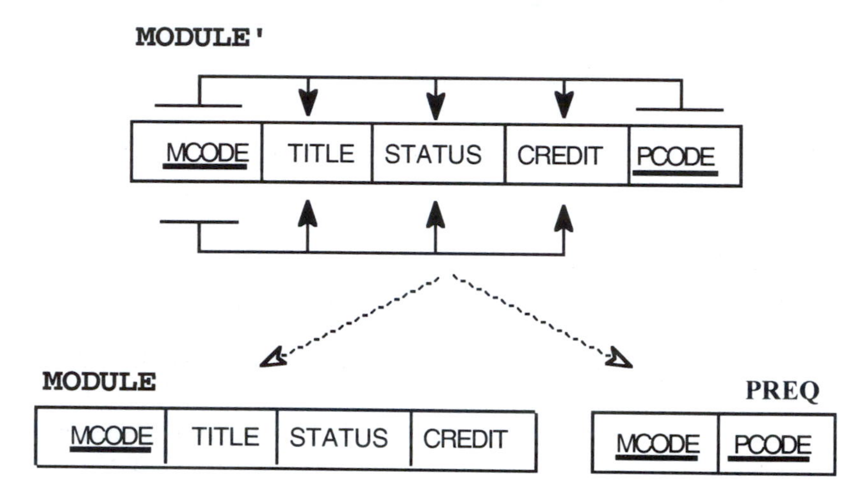

Fig 11.2

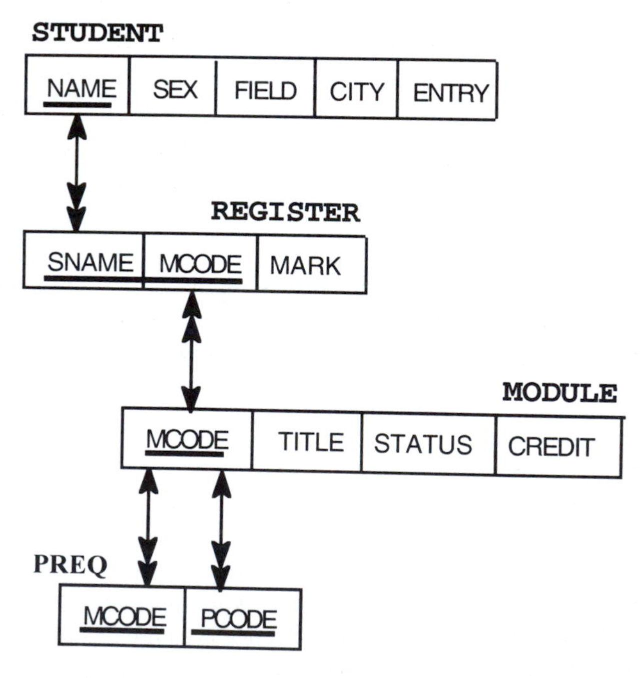

Fig. 11.3

The final structure for the database is shown in Fig 11.3. Note that the primary key in PREQ is (MCODE, PCODE) and hence neither attribute can be null. Consequently, the modules with no prerequisites will be referenced to an artificially introduced dummy module (0, 0)

11.2 SOFTWARE INSTALLATION

The development of the application characterized above was carried out using the following software:

<table>
<tr><td>Operating System</td><td>Linux 2.2</td><td>Redhat 6.0</td></tr>
<tr><td>Database System</td><td>MySQL 3.23</td><td rowspan="3">Additional modules:
DBI 1.13
Msql-Mysql-modules-1.2215</td></tr>
<tr><td>Application Code</td><td>Perl 5.005_03</td></tr>
<tr><td>Web Server</td><td>Apache 1.3.9</td></tr>
<tr><td>Web Browser</td><td>Any can be used</td><td>Lynx, Mozilla and Netscape are free and run on Linux</td></tr>
</table>

The products used in this Case Study can be freely downloaded over the Internet (for non-commercial use, in any case). This book's companion website contains the relevant links to the appropriate locations, as well as reasonably comprehensive installation instructions and, of course, all of the SQL and Perl code used throughout this chapter. It has to be said, however, that software installation on a PC, whilst interesting and educational, is not necessarily a trivial exercise.

MySQL is a small but fast relational DBMS. What it lacks in functionality (most noticeably, neither foreign keys nor data constraints are supported) is made up for in its performance, simplicity and free licence.

Perl is a very flexible scripting language, popular in the Unix world for quite some time, and a prevalent choice for CGI implementations. Despite it being an interpreted language, it offers a very good performance; moreover many modules are available to simplify many tasks.

Apache is one of the most popular web servers in use on the Internet (see, for instance http://www.netcraft.com/survey), and offers a responsive and high performance platform which integrates well with other technologies.

DBI provides abstracted database function calls to enable Perl to execute commands in the chosen database system. Each database system has its separate DBD module, which links the abstracted interface to the native client library for the particular database.

We would like to emphasize that this Chapter is not intended to serve as a description of the software packages involved; the objective is to give a simple introduction, using a simple example, of how various components can be linked together to form an application. For detailed description of these packages the reader is advised to consult Schwartz (1997), Laurie (1999), Yarger (1999) and Wall (2000).

11.3 IMPLEMENTATION

The simplest way of demonstrating how the components can be linked together is shown below:

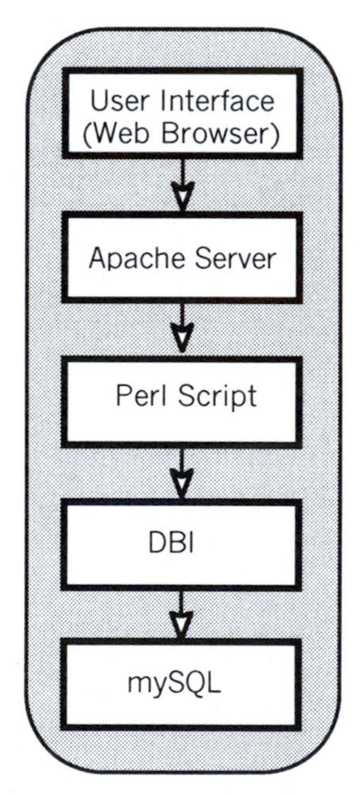

Fig. 11.4

Running a web browser against the application URL, such as:

http://mymachine.somedomain.co.uk/cgi-bin/list-students.pl

will cause a request to be sent to the Apache web server. Apache will then interpret the call, execute the appropriate Perl program (say *list-students.pl*), and pass in any data inserted into the relevant forms.

The program *list-students.pl*, acting on the data passed in, will open the database connection and execute the relevant SQL command. The returned data is then formatted in HTML and streamed back, through the web server, to the client browser.

The database tables as described in the first section of this Chapter were created by standard SQL definitions; their structures are shown in Figs 11.5 - 11.8.

```
mysql> desc STUDENT;
+-------+----------+------+-----+---------+-------+
| Field | Type     | Null | Key | Default | Extra |
+-------+----------+------+-----+---------+-------+
| NAME  | char(8)  |      | PRI |         |       |
| SEX   | char(1)  | YES  |     | NULL    |       |
| FIELD | char(4)  |      |     |         |       |
| CITY  | char(15) | YES  |     | NULL    |       |
| ENTRY | date     | YES  |     | NULL    |       |
+-------+----------+------+-----+---------+-------+
5 rows in set (0.00 sec)
```

Fig. 11.5

```
mysql> desc MODULE;
+--------+----------+------+-----+---------+-------+
| Field  | Type     | Null | Key | Default | Extra |
+--------+----------+------+-----+---------+-------+
| CODE   | char(4)  |      | PRI |         |       |
| TITLE  | char(15) | YES  |     | NULL    |       |
| STATUS | char(1)  | YES  |     | NULL    |       |
| CREDIT | int(11)  | YES  |     | NULL    |       |
+--------+----------+------+-----+---------+-------+
4 rows in set (0.00 sec)
```

Fig. 11.6

```
mysql> desc REGISTER;
+-------+---------+------+-----+---------+-------+
| Field | Type    | Null | Key | Default | Extra |
+-------+---------+------+-----+---------+-------+
| MCODE | char(4) |      | PRI |         |       |
| SNAME | char(8) |      | PRI |         |       |
| MARK  | int(3)  | YES  |     | NULL    |       |
+-------+---------+------+-----+---------+-------+
3 rows in set (0.00 sec)
```

Fig. 11.7

```
mysql> desc PREQ;
+-------+---------+------+-----+---------+-------+
| Field | Type    | Null | Key | Default | Extra |
+-------+---------+------+-----+---------+-------+
| MCODE | char(4) |      | PRI |         |       |
| PCODE | char(4) |      | PRI |         |       |
+-------+---------+------+-----+---------+-------+
2 rows in set (0.00 sec)
```

Fig. 11.8

The application presented in this Chapter consists of the following five processes:

- Reporting on a student's module enrolment (a simple retrieval)
- Adding a student's enrolment (a simple insert operation)
- Deleting a student's enrolment on a module (a simple delete operation)
- Deleting a module from the database (a more complex delete operation)
- Adding a module to the database (a more complex insert operation).

The application code was developed with the focus on the integration of components rather than appearance of the results. While the attractiveness of user interfaces is important in application systems, user interface design (whether by using HTML or dedicated software packages) is not within the scope of this book.

Also, the Perl programs included in this Chapter are intended for explanatory purposes rather than being examples of professional coding; issues such as code optimization, whether from the performance or structural standpoints, are not considered at all.

11.3.1 Student enrolment report

A simple SQL statement to report on a student's module enrolment set against the above database schema can be written as:

```
select MODULE.CODE, MODULE.TITLE, STUDENT.NAME
from STUDENT, REGISTER, MODULE
where STUDENT.NAME ="John"
and STUDENT.NAME = REGISTER.SNAME
and REGISTER.MCODE = MODULE.CODE
```

The task now is to implement this action within the Perl environment. The database interface (DBI) allows SQL queries to be passed directly to the underlying database system. On execution of the select query, the DBI provides a data structure which is then processed, as necessary. The following code demonstrates how this is implemented:

```
1    #!/usr/bin/perl
2
3    ###
4    # Define which perl modules we need
5    ###
6    use strict;
7    use CGI qw(:standard);
8    use DBI();
9
10   ### Create a CGI handler
11   my ($output)=new CGI;
12
13   # Connect to the database.
14   # the $dbh variable then becomes a handle into the database.
15
16   my $dbh = DBI->connect("DBI:mysql:database=test;host=localhost",
17                "mysql", "",
18        {'RaiseError' => 1});
19
20   # start the html correctly
21   print header,start_html;
22
23   ### Define the SQL Statement to be executed.
24   my $sth = $dbh->prepare("\
25   SELECT
     MODULE.CODE,MODULE.TITLE,STUDENT.NAME,REGISTER.MARK \
26     FROM STUDENT,REGISTER,MODULE \
27    WHERE STUDENT.NAME=\"John\"  \
28      AND STUDENT.NAME=REGISTER.SNAME   \
```

```
29        AND REGISTER.MCODE=MODULE.CODE ");
30
31    ###
32    # Execute the statement
33    ###
34    $sth->execute();
35
36    print "<PRE>John's modules:<br>";
37    print "Code\tTitle\t\tGrade<br>";
38
39    ###
40    # Whilst there are rows to return, print them out
41    ###
42    while (my $ref = $sth->fetchrow_hashref()) {
43            print "$ref->{'CODE'}\t$ref->{'TITLE'}\t$ref->{'MARK'}<br>";
44    }
45
46    $sth->finish;
47
48    # Tidy up
49    print "</PRE>";
50    print end_html;
51
52    # and disconnect.
53    $dbh->disconnect;
```

Line 16 initiates a connection with the database, and gives the controlling Perl program a handle ($dbh) to refer to at a later date. The program could have multiple connections to the same, or other databases, not all of them being necessarily mySQL.

Lines 11, 21, and 48-50 deal with outputting the necessary control codes for the HTML environment in which the Perl program is run; without these the browser would not be able to interpret the returned results. Note that, for simplicity, the <PRE> tag defines the text as "pre-formatted", and the browser can display it in it's raw fashion. However, header and footer text are necessary to ensure the correct character sets are shown.

Lines 24-29 define the SQL statement. Note that the double quotes have to be \"escaped"\ to avoid confusion with the double quotes used to define the boundaries of the SQL code. This results in a control handle being returned. $dbh->prepare calls a function internally to the database handle created earlier.

Execution of the statement takes place on line 34, but no processing of the results occurs yet for the program needs to activate the control handle for the SQL query row by row. This occurs within the while loop defined on line 42, which cycles until there is no more data to be returned. Each line inside the while loop prints a particular field from the resulting data.

The above program, when placed into an appropriate directory on a web server, can then be called through a browser:

    http://some.domain.name/cgi-bin/show-enrol.pl

The web server opens and executes the program, passing back to the user's browser the output from the program, which is a formatted HTML listing of John's modules:

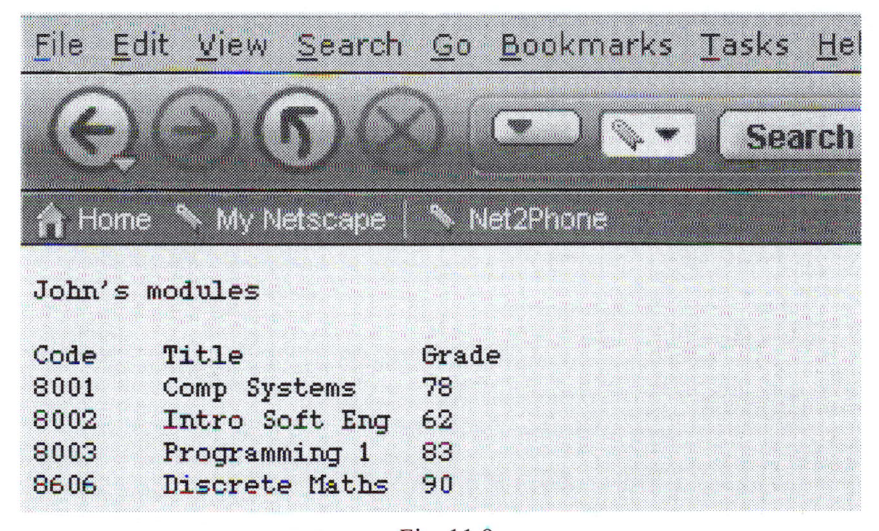

Fig. 11.9

The above example is both very simple and not particularly flexible; it does not even allow the user to input any value for the attribute NAME. The next example will address this point.

11.3.2 Insert student enrolment

This program inserts a new record into the table REGISTER to store a student's result for a given module. Essentially, the programme is composed of two parts. First, the condition (line 22) checks whether there are any parameters in the invocation of the program. If there are (lines 33-37), then the "add_enrolment" subroutine is called, otherwise (lines 22-32) the available students and modules are displayed (by calling the appropriate subroutines in lines 118-156).

The section between the lines 46 and 113 expresses the data validation and, consequently, insertion of the record. The code would have been significantly simpler if MySQL supported foreign keys, data constraints and triggers, in which case almost entire validation would have taken place inside the database engine.

However, as this is not the case, the application code must implement the relevant controls - for each data item to be inserted into REGISTER, the value or existence in referenced tables need to be verified. Should any part of verification fail, (for example, the module does not exist), the error message is displayed and exit from the subroutine activated.

```
1       #!/usr/bin/perl
2
3       ###
4       # Define which perl modules we need
5       ###
6       use strict;
7       use CGI qw(:standard);
8       use DBI();
9
10      ### Create a CGI handler
11      my ($output)=new CGI;
12
13      # Connect to the database.
14      # the $dbh variable then becomes a handle into the database.
15
16      my $dbh = DBI >connect("DBI:mysql:database=test;host=localhost",
17                  "mysql", "",
18          {'RaiseError' => 1});
19
20      print header,start_html;
21
22      if (not $output->param) {
23                # start the html correctly
24                print "<PRE>";
25
26                print "Defined Students:\n";
27                &show_students;
28
29                print "\n\nDefined Modules:\n";
30                &show_modules;
```

```
31
32              &display_form;
33      } else {
34              print "<PRE>";
35              print "Checking...<br>";
36              &add_enrolment($output->param('mod_code'),$output-
                >param('stu_name'),$output->param('mark'));
37      }
38
39
40      # Tidy up
41      print end_html;
42
43      # and disconnect.
44      $dbh->disconnect;
45
46      ###
47      # Add a student result
48      ###
49      sub add_enrolment {
50              my($mcode,$sname,$mark)=@_;
51              my($norows);
52
53              ### validation checks
54              if (!( ($mark>=0) && ($mark<=100) )) {
55                print("<PRE>Mark Not Valid (Must be 0-100)\n</PRE>");
56                      exit;
57              }
58
59              ###
60              # Is it a valid module code?
61              ###
62              my $sth=$dbh->prepare("SELECT CODE FROM MODULE
                                        WHERE CODE=\"$mcode\"");
63              $sth->execute;
64
65              # See how many rows were returned.
66              $norows=$sth->rows;
67              $sth->finish;
68
69              if ($norows==0) {
70                      print("Invalid module ($mcode) -
                                    Choose from the following:\n\n");
71                      &show_modules;
72                      exit;
73              }
74
```

```
 75          ###
 76          # Does the student exist?
 77          ###
 78          my $sth=$dbh->prepare("SELECT NAME FROM STUDENT
                                 WHERE NAME=\"$sname\"");
 79          $sth->execute;
 80          $norows=$sth->rows;
 81          $sth->finish;
 82
 83          if ($norows==0) {
 84                  print("Invalid student ($sname) -
                                  Choose from the following:\n\n");
 85                  &show_students;
 86                  exit;
 87          }
 88
 89          ###
 90          # Do we already have a record!?
 91          ###
 92          my $sth=$dbh->prepare
                 ("SELECT MCODE,SNAME,MARK FROM REGISTER
                 WHERE MCODE=\"$mcode\" AND SNAME=\"$sname\"");
 93          $sth->execute;
 94          $norows=$sth->rows;
 95          if ($norows==1) {
 96           print("A record already exists for this module/student: \n");
 97           while (my $ref = $sth->fetchrow_hashref()) {
 98                  print "Module: $ref->{'MCODE'}\nStudent:
                     $ref->{'SNAME'}\nMark: $ref->{'MARK'}<br>";
 99                  }
100                  $sth->finish;
101                  exit;
102          }
103
104          ###
105          # Now add the record...
106          ###
107
108          my $sth=$dbh->do("INSERT INTO REGISTER
             VALUES(?,?,?)",undef,$mcode,$sname,$mark) || print("An
             error occured adding mark: $dbh->errstr\n");
109
110          $sth->finish;
111
112          print("Record successfully submitted!\n");
113    }
114
```

```
115     ###
116     # List the available students
117     ###
118     sub show_students {
119
120          my $sth = $dbh->prepare("SELECT NAME FROM STUDENT");
121          $sth->execute;
122

123          while (my $ref = $sth->fetchrow_hashref()) {
124                       print "$ref->{'NAME'}<br>";
125          }
126     }
127
128     ###
129     # List the available modules
130     ###
131     sub show_modules {
132
133         my $sth = $dbh->prepare("SELECT CODE,TITLE FROM MODULE");
134         $sth->execute;
135
136         while (my $ref = $sth->fetchrow_hashref()) {
137                       print "$ref->{'CODE'}\t$ref->{'TITLE'}<br>";
138         }
139         $sth->finish;
140     }
14
142     sub display_form {
143              ###
144              # Build the form
145              ###
146              print <<FORMEND;
147     </PRE>
148     <H2>Enter Module Code, Student Name and Mark:</H2>
149     <FORM METHOD=post ACTION="add-enrol.pl">
150     code: <INPUT NAME="mod_code" SIZE=4>
151     name: <INPUT NAME="stu_name" SIZE=8>
152     mark: <INPUT NAME="mark" SIZE=3>
153     <INPUT TYPE=SUBMIT VALUE="Submit" HIDDEN NAME="action">
154     FORMEND
155
156     }
```

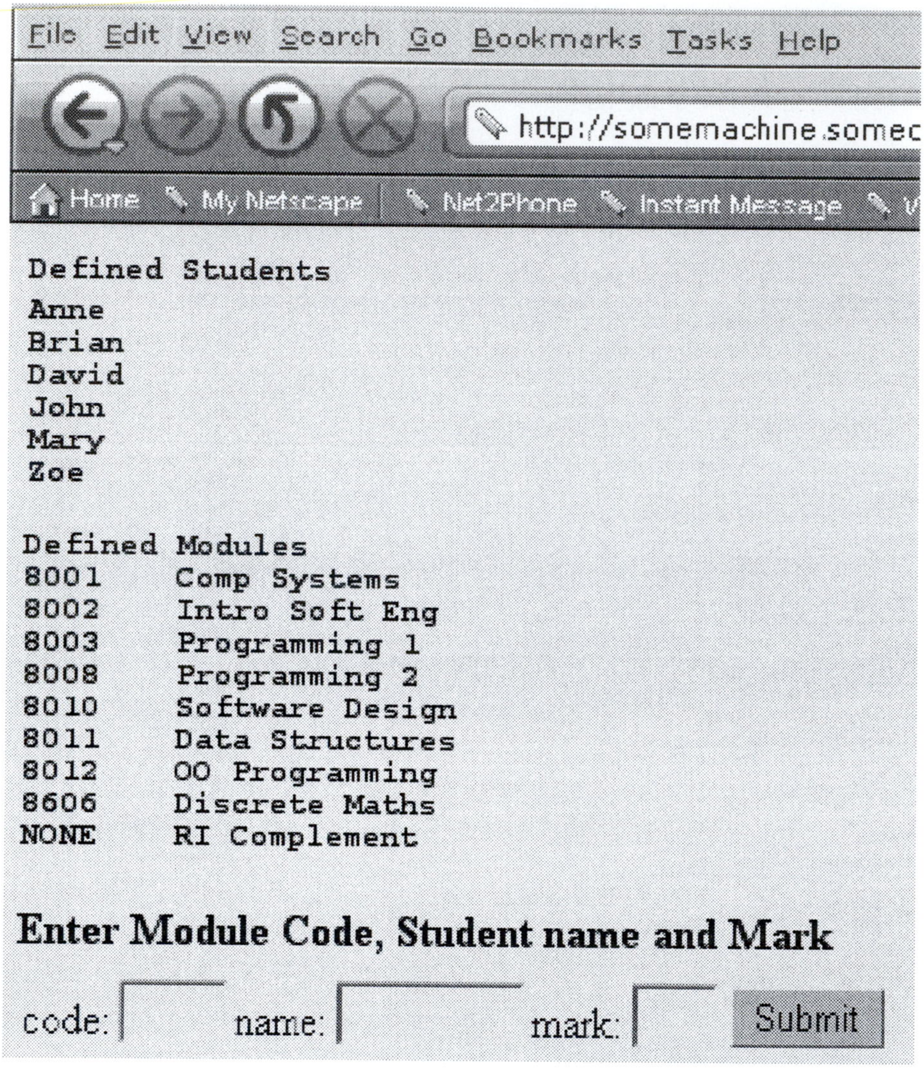

File Edit View Search Go Bookmarks Tasks Help

http://somemachine.somec

Home My Netscape Net2Phone Instant Message

```
Defined Students
Anne
Brian
David
John
Mary
Zoe

Defined Modules
8001    Comp Systems
8002    Intro Soft Eng
8003    Programming 1
8008    Programming 2
8010    Software Design
8011    Data Structures
8012    OO Programming
8606    Discrete Maths
NONE    RI Complement
```

Enter Module Code, Student name and Mark

code: name: mark: Submit

Fig. 11.10

The two examples so far give a simple demonstration on how a relatively simple Perl script can connect to the database and process the results. The second one demonstrates how the Perl program, built as a CGI script, can interface with a web browser and a server thus producing a system that can be used over a network. It is also possible, as demonstrated in the insert example, to build systems that could even provide support for those functions which a DBMS would normally be expected to operate (i.e. the database integrity management).

11.3.3 Delete student enrolment

In the interest of space, only the code of the subroutine which would perform the deletion is presented rather than the full program. This subroutine checks that the parameters have been passed correctly (lines 6-9) and, in a manner similar to that in the insertion example, checks that the given student/module combination exists before attempting to delete the record (lines 12-17); lines 22-24 perform the actual deletion and return the termination message.

```
 1 sub delete_enrolment {
 2        my ($mcode,$sname)=@_;
 3        my ($norows);
 4
 5        # Were the parameters properly populated?
 6       if ( (!($mcode)) || (!($sname)) ) {
 7               print("Please fill in the form properly!\n");
 8               exit;
 9       }
10
11        # check enrolment
12         my $sth=$dbh->prepare
              ("select SNAME,MCODE from REGISTER where
                SNAME=\"$sname\" and MCODE=\"$mcode\"");
13        $sth->execute;
14         $norows=$sth->rows;
15
16        # There were no rows found in the check!
17        if ($norows==0) {
18           # Display error
19           print
             ("Student [$sname] is NOT enrolled on module [$mcode]\n");
20           exit;
21        } else {
22               # Delete the record.
23               my $sth=$dbh->do("delete from REGISTER where
                        SNAME=\"$sname\" and MCODE=\"$mcode\"");
24              print("Student [$sname] results have been removed from
                        module [$mcode]\n");
25       }
26 }
```

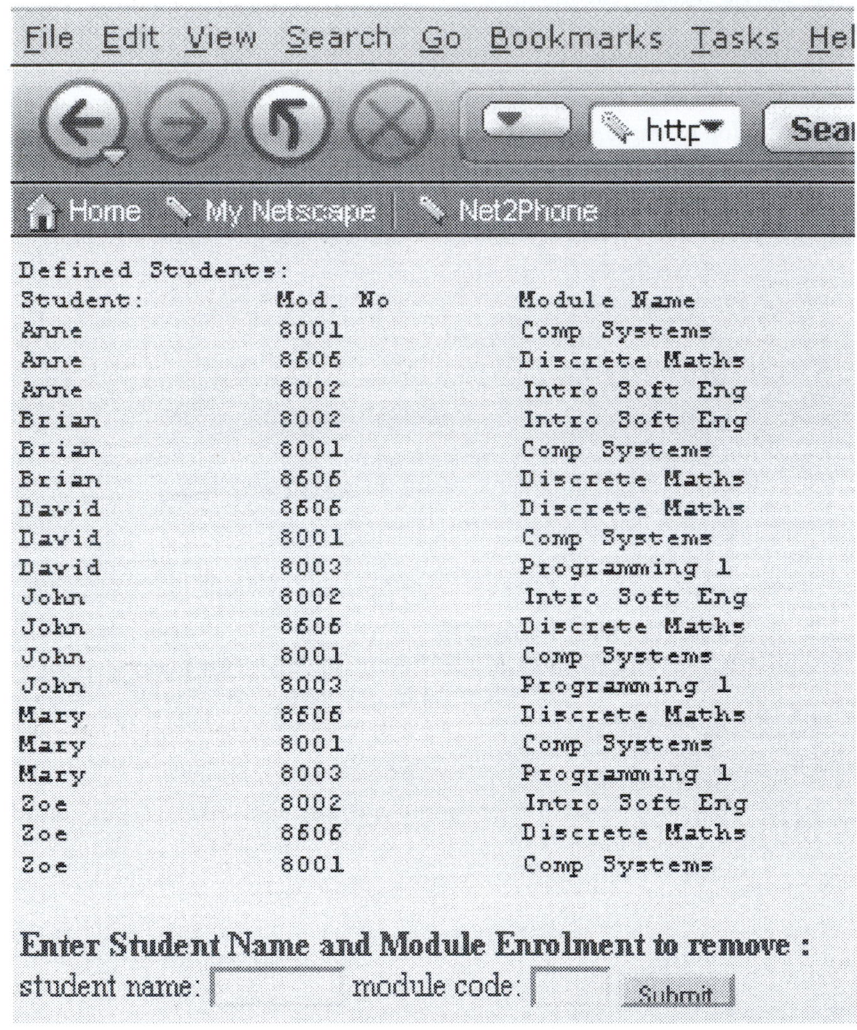

Fig. 11.11

11.3.4 Delete a module from the database

A more complex delete operation is a transaction that could potentially violate some constraints. For example, the deletion of a module record in MODULE could invalidate the data held in REGISTER, and possibly affect the PREQ table, forcing redefinition of the pre-requisite structure.

The core of the program is the *delete_module* subroutine at line 139, which firstly checks that the module does not form a pre-requisite to another module, and that there are no students enrolled on the module. Only when the database consistency has been validated does the program proceed to delete the record. Likewise, a program that maintained module dependencies would need to ensure that in adding a module dependency to the system, the newly inserted module is not linked to a module that does not exist.

```
1    #!/usr/bin/perl
2
3    ###
4    # Define which perl modules we need
5    ###
6    use strict;
7    use CGI qw(:standard);
8    use DBI();
9
10   ### Create a CGI handler
11   my ($output)=new CGI;
12
13   # Connect to the database.
14   # the $dbh variable then becomes a handle into the database.
15
16   my $dbh = DBI>connect("DBI:mysql:database=test;host=localhost",
17             "mysql", "",
18         {'RaiseError' => 1});
19
20   print header,start_html;
21
22   my $nparams=$output->param;
23
24   if (!($output->param)) {
25             # start the html correctly
26             print "<PRE>";
27
28             print "Defined Modules:\n";
29             &show_modules;
30
31             &display_form;
32   } else {
33             print "<PRE>";
34             &delete_module($output->param('mcode'));
35   }
36
37
```

```
38    # Tidy up
39    print end_html;
40
41    # and disconnect.
42    $dbh->disconnect;
43
44
45    sub display_form {
46              ###
47              # Build the form
48              ###
49              print <<FORMEND;
50    </PRE>
51    <H2>Enter Module Code to remove :</H2>
52    <FORM METHOD=post ACTION="delete-module.pl">
53    module: <INPUT NAME="mcode" SIZE=4>
54    <INPUT TYPE=SUBMIT VALUE="Submit" HIDDEN NAME="action">
55    FORMEND
56
57    }
58
59    sub display_enrolment {
60
61              my $sth = $dbh->prepare("select SNAME,CODE,TITLE    \
62                        from STUDENT, REGISTER,MODULE \
63                        where STUDENT.NAME=REGISTER.SNAME \
64                        and REGISTER.MCODE=MODULE.CODE \
65                        order by SNAME
66              ");
67              $sth->execute;
68
69              print("Student:\tMod. No\t\tModule Name\n");
70              while (my $ref = $sth->fetchrow_hashref()) {
71              print "$ref->{'SNAME'}\t\t$ref->{'CODE'}\t\t$ref >{'TITLE'}<br>";
72              }
73
74    }
75
76    ###
77    # List the available modules
78    ###
79    sub show_modules {
80              my ($nostudents);
81
82              my $sth = $dbh->prepare
                                  ("SELECT CODE,TITLE FROM MODULE");
83              $sth->execute;
```

```
84
85            while (my $ref = $sth->fetchrow_hashref()) {
86              print "$ref->{'CODE'}\t$ref->{'TITLE'}<br>";
87              show_prereqs($ref->{'CODE'});
88              $nostudents=&check_enrol($ref->{'CODE'});
89              print("There are $nostudents students enrolled on the module\n");
90              print "\n";
91            }
92            $sth->finish;
93    }
94
95    ###
96    # Show the given modules pre-reqs
97    ###
98    sub show_prereqs {
99            my($mcode)=@_;
100           my($norows);
101
102           my $sth=$dbh->prepare("SELECT CODE,TITLE \
103                          FROM MODULE,PREQ \
104                          WHERE MODULE.CODE=PREQ.MCODE \
105                          AND PREQ.PCODE=\"$mcode\"
                             AND PREQ.PCODE!=\"NONE\" ");
106           $sth->execute;
107           $norows=$sth->rows;
108
109           if ($norows>0) {
110                   while (my $ref = $sth->fetchrow_hashref()) {
111                          print "PRE-REQUISITE TO $mcode:
                             $ref->{'CODE'}\t$ref->{'TITLE'}<br>";
112                   }
113           } else {
114                   print("NO PRE-REQUISITES\n");
115           }
116           $sth->finish;
117   }
118

119   ###
120   # Check pre-reqs
121   ###
122   sub check_prereq {
123           my ($mcode)=@_;
124           my ($norows);
125
126           # Check if the module is a pre-req for another.
127           my $sth=$dbh->prepare
```

```
                ("SELECT MCODE FROM PREQ WHERE PCODE=\"$mcode\"");
128             $sth->execute;
129             $norows=$sth->rows;
130             $sth->finish;
131
132             # return number of modules it is a pre-req to
133             return ($norows);
134     }
135
136     ###
137     # delete a module
138     ###
139     sub delete_module {
140
141             my($mcode)=@_;
142
143             if (&check_prereq($mcode)>0) {
144                     print("Module [$mcode] forms a pre-requisite!\n");
145                     &show_prereqs($mcode);
146                     exit;
147             }
148
149             if (&check_enrol($mcode)>0) {
150                     print("Module [$mcode] has students enrolled!\n");
151                     exit;
152             }
153
154             print("OK to delete\n");
155             # delete modules.
156             my $sth=$dbh->do
                ("DELETE FROM MODULE WHERE CODE=\"$mcode\"");
157
158             # delete the modules pre-requisites.
159             my $sth=$dbh->do
                    ("DELETE FROM PREQ WHERE MCODE=\"mcode\"");
160     }
161
162
163     ###
164     # Check to see if any students are enrolled!
165     ###
166     sub check_enrol {
167             my ($mcode)=@_;
168             my ($norows);
169
170             # Check if the module is a pre-req for another.
171             my $sth=$dbh->prepare
```

```
                ("SELECT MCODE FROM REGISTER
                WHERE MCODE=\"$mcode\"");
172             $sth->execute;
173             $norows=$sth->rows;
174             $sth->finish;
175
176             # return number of modules it is a pre-req to
177             return ($norows);
178   }
```

Another significant difference with regard to the earlier programs is the shift of certain checks to subroutines, in the standard way promoted by structured programming. Thus, the subroutines can be called at multiple points within the program.

Needless to say, this could become a performance bottleneck if repeated excessively. A solution, however, might be to cache details such as module pre-requisites at the invocation of the program, thereby reducing the number of times the database would be hit.

Also, there is no reason why, following the deletion of a student's enrolment on a particular module, the system could not check to see whether the deletion has removed all remaining results from a module, and could suggest to the user to remove the module record. However, it is important to remember that every check made by a program that results in a query against the database consumes resources on the system. In a large, high-capacity database system, additional or repetitive checks could easily become a significant performance problem and thus a rational compromise is inevitable.

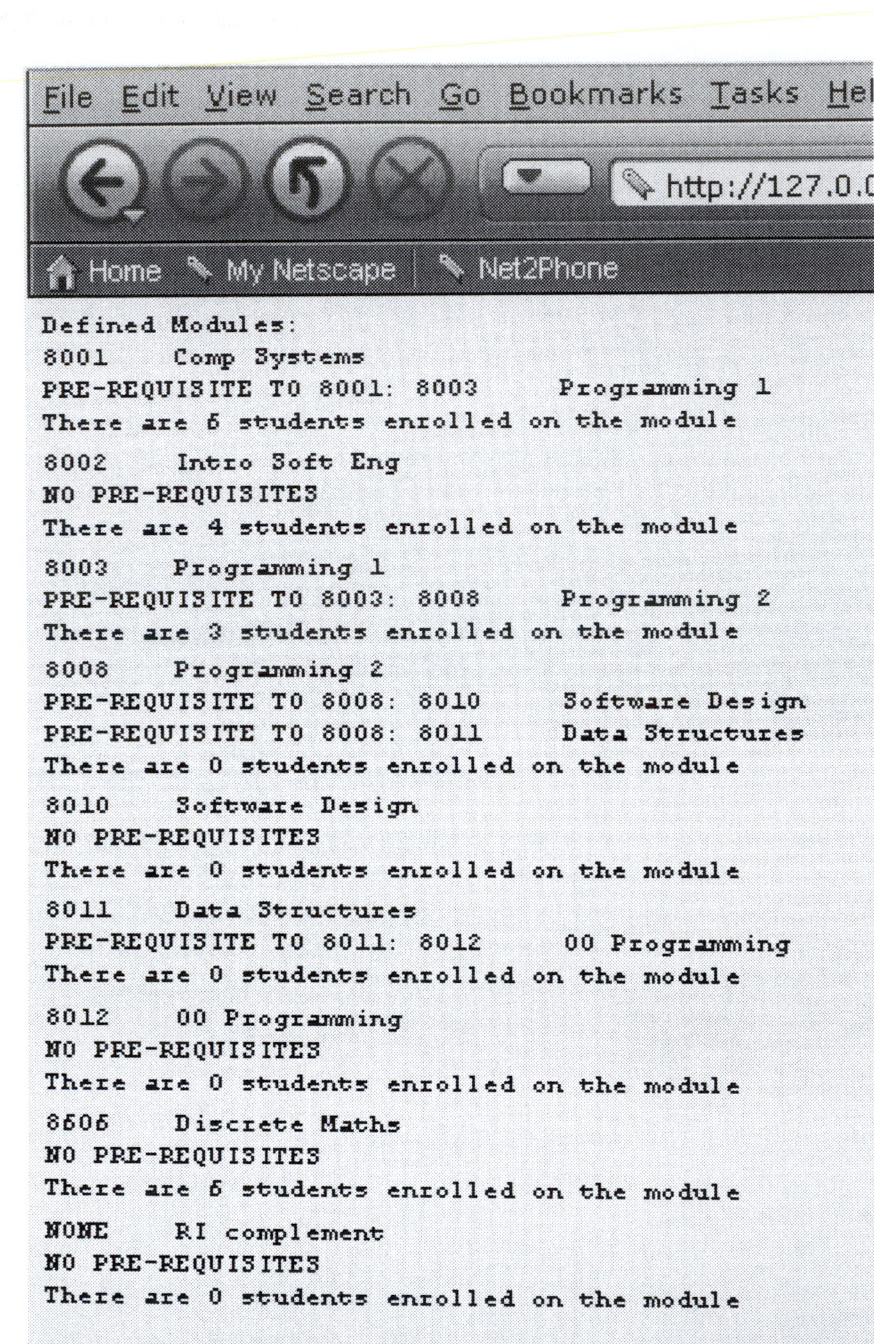
File Edit View Search Go Bookmarks Tasks Hel
http://127.0.0
Home My Netscape Net2Phone
Defined Modules:
8001 Comp Systems
PRE-REQUISITE TO 8001: 8003 Programming 1
There are 6 students enrolled on the module
8002 Intro Soft Eng
NO PRE-REQUISITES
There are 4 students enrolled on the module
8003 Programming 1
PRE-REQUISITE TO 8003: 8008 Programming 2
There are 3 students enrolled on the module
8008 Programming 2
PRE-REQUISITE TO 8008: 8010 Software Design
PRE-REQUISITE TO 8008: 8011 Data Structures
There are 0 students enrolled on the module
8010 Software Design
NO PRE-REQUISITES
There are 0 students enrolled on the module
8011 Data Structures
PRE-REQUISITE TO 8011: 8012 OO Programming
There are 0 students enrolled on the module
8012 OO Programming
NO PRE-REQUISITES
There are 0 students enrolled on the module
8606 Discrete Maths
NO PRE-REQUISITES
There are 6 students enrolled on the module
NONE RI complement
NO PRE-REQUISITES
There are 0 students enrolled on the module
Enter Module Code to remove :
module:
Submit

Fig. 11.12

11.3.5 Add a module to the database

Finally, we deal with the addition of a module to the database. Since a module may have one or more pre-requisites their existence in the database must be verified during the insertion process.

```
 1 #!/usr/bin/perl
 2
 3 ###
 4 # Define which perl modules we need
 5 ###
 6 use strict;
 7 use CGI qw(:standard);
 8 use DBI();
 9
10 ### Create a CGI handler
11 my ($output)=new CGI;
12
13 # Connect to the database.
14 # the $dbh variable then becomes a handle into the database.
15
16 my $dbh = DBI->connect("DBI:mysql:database=test;host=localhost",
17          "mysql", "",
18         {'RaiseError' => 1});
19
20 print header,start_html;
21
22 my $nparams=$output->param;
23
24 if (!($output->param)) {
25         # start the html correctly
26         print "<PRE>";
27
28         print "Currently Defined Modules:\n";
29         &show_modules;
30
31         &display_form;
32 } else {
33         print "<PRE>";
34         &insert_module($output->param('mcode'),
                                                  $output >param('title'),
35                        $output->param('status'),$output->param('credit'),
36                        $output->param('depend'));
37 }
38
```

```
39
40  # Tidy up
41  print end_html;
42
43  # and disconnect.
44  $dbh->disconnect;
45
46
47  sub display_form {
48          ###
49          # Build the form
50          ###
51          print <<FORMEND;
52  </PRE>
53      <B>Enter New Module details:</B>
54      <FORM METHOD=post ACTION="add-module.pl">
55    Module code: <INPUT NAME="mcode" SIZE=4>
56    Module title: <INPUT NAME="title" SIZE=20> <br><br>
57    Module status: <INPUT NAME="status" SIZE=1>
58    Module credit: <INPUT NAME="credit" SIZE=5><br><br>
59    Module dependency: <INPUT NAME="depend" SIZE=4>
60
61  <INPUT TYPE=SUBMIT VALUE="Submit" HIDDEN NAME="action">
62  FORMEND
63
64  }
65
66  ###
67  # List the available modules
68  ###
69  sub show_modules {
70        my ($nostudents);
71
72          my $sth = $dbh->prepare("SELECT CODE,TITLE FROMMODULE");
73        $sth->execute;
74
75        while (my $ref = $sth->fetchrow_hashref()) {
76                  print "$ref->{'CODE'}\t$ref->{'TITLE'}<br>";
77        }
78        $sth->finish;
79  }
80
81  ###
82  # Check the specified module exists
83  ###
```

```
84  sub check_exists {
85          my ($mcode)=@_;
86          my ($norows);
87
88          # Check if the module appears in MODULE
89           my $sth=$dbh->prepare("SELECT CODE FROM MODULE
                                     WHERE CODE=\"$mcode\"");
90          $sth->execute;
91           $norows=$sth->rows;
92          $sth->finish;
93
94          return ($norows);
95  }
96
97  ###
98  # insert a module
99  ###
100  sub insert_module {
101
102          my($mcode,$mtitle,$mstatus,$mcredit,$mdepend)=@_;
103
104          ## firstly, is the module code already in use?
105          if (&check_exists($mcode)>0) {
106                  print("Module code [$mcode] already used!\n");
107                  exit;
108          }
109
110          ### if a dependency was specified
111          if ($mdepend) {
112                  ### does that dependency exist?
113                  if (&check_exists($mdepend)==0) {
114                          print("Invalid dependency. Dependent Module
                                  [$mdepend] does NOT exist!\n");
115                          exit;
116                  }
117          }
118
119          ### insert the module record
120
121           my $sth=$dbh->do("INSERT INTO MODULE VALUES
         (?,?,?,?)",undef,$mcode,$mtitle,$mstatus,$mcredit) || print
         ("An error occured adding module: $dbh->errstr\n");
122
123          ### insert the dependency record, if appropriate
124          if ($mdepend) {
```

```
125            my $sth=$dbh->do("INSERT INTO PREQ VALUES
              (?,?)",undef,$mcode,$mdepend) || print ("An error
               occured adding module dependency: $dbh->errstr\n");
126        } else {
127            my $sth=$dbh->do("INSERT INTO PREQ VALUES
              (?,?)",undef,$mcode,"NONE") || print ("An error occured
              adding module dependency: $dbh->errstr\n");
128        }
129 }
```

File Edit View Search Go Bookmarks Tasks Help

http://son Search

Home My Netscape Net2Phone Instant Message

```
Currently Defined Modules:
8001    Comp Systems
8002    Intro Soft Eng
8003    Programming 1
8008    Programming 2
8010    Software Design
8011    Data Structures
8012    OO Programming
8606    Discrete Maths
NONE    RI complement
```

Enter New Module details:

Module code: 8050 Module title: Perl Programming

Module status: A Module credit: 1

Module dependency: 8008 Submit

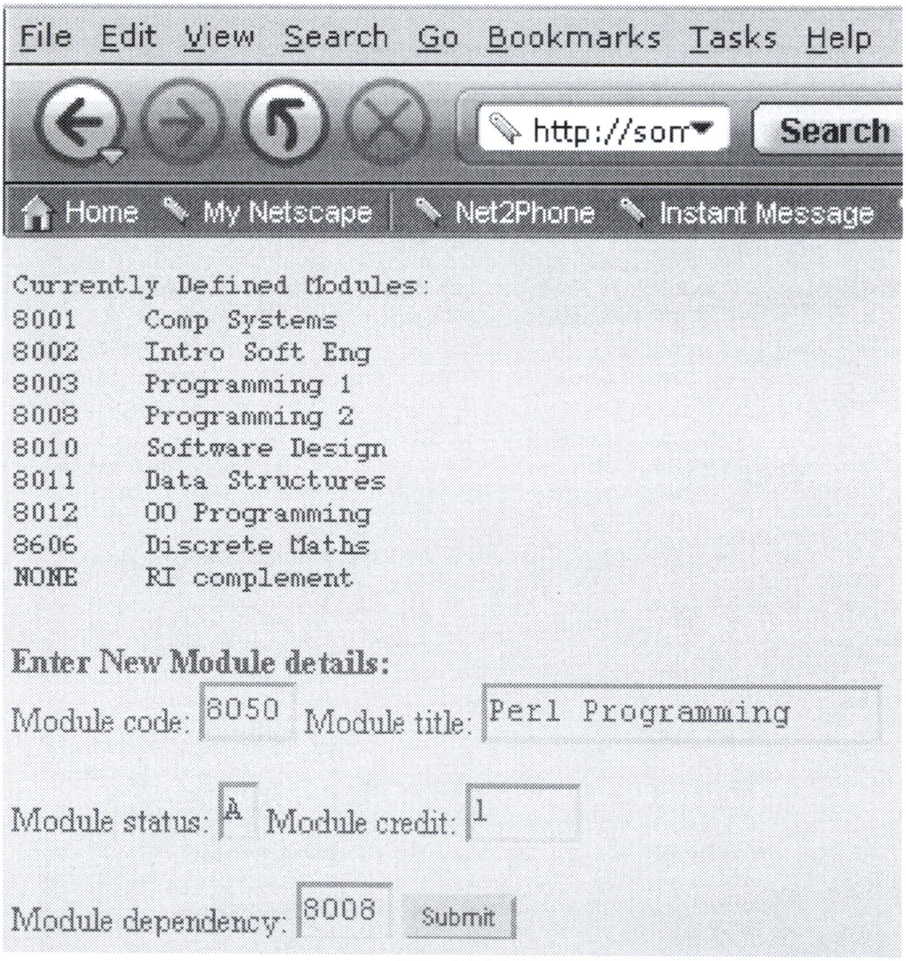

Fig. 11.13

The subroutine *check_exists* (line 84) verifies that a given module exists. This is then used in the core of the program (the subroutine *insert_module* at line 100) to verify that the module insertion does not violate the primary key in MODULE (line 105), but also to check the existence of the pre-requisite module (line 111). The insertion occurs between lines 124 and 128; however even if the new module has no pre-requisite, the 'complementary' module "NONE" serves that purpose.

It is worth noting that if the foreign key constraint was enforced through the data definition, then the core of the *check_exists* subroutine would be given as:

```
my $sth=$dbh->prepare("SELECT CODE FROM PREQ WHERE
                                    MCODE=\"$mcode\"");
```

We have presented this very simple implementation to show how the database would fit in as a core component of an application system. We have also demonstrated that, had the underlying DBMS not supported integrity constraints (in particular those resulting from the foreign keys), these constraints could nevertheless be imposed by other means, though yielding the application code of much increased complexity.

Clearly, the combination of Perl, MySQL and Apache may not be appropriate in all circumstances. There are various alternatives and approaches by which the underlying database could be incorporated into the overall application design. It is, therefore, important to consider as early as possible in the development life cycle the questions of which software products to choose and how their components are linked together.

11.4 EXERCISES

11.1 Download and install the software mentioned in 11.2. Install the examples (available from the book's website) and extend the application to support other information needs of the potential users.

11.2 Investigate alternative DBD interfaces to database systems you have access to. How could the program be improved through use of the DBMS-specific features (such as those discussed in Chapter 10 on extended SQL)?

11.3 What benefits or problems could be introduced by removing the database abstraction layer (DBI/DBD) and using native DBMS function libraries?

Appendix A: Solutions to exercises

CHAPTER 2

2.1 Examples of assumptions are: existence of some entities (e.g. COMPUTER SYSTEM); existence, degrees and participation conditions of some relationships (e.g. between LECTURER and STUDENT)

2.2

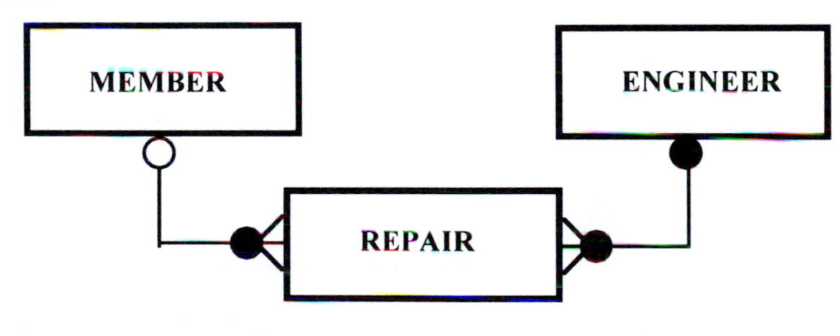

2.3

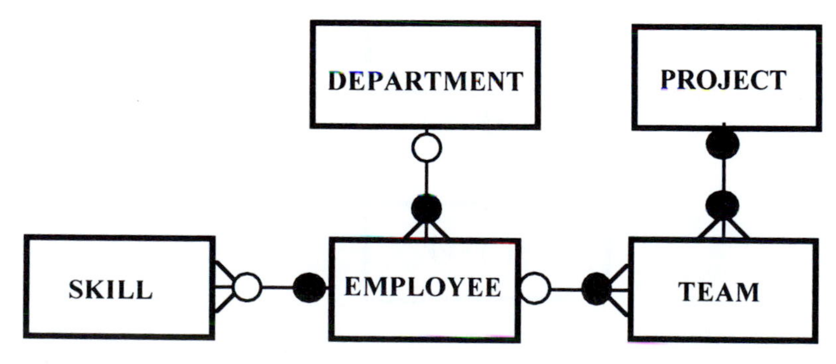

2.4

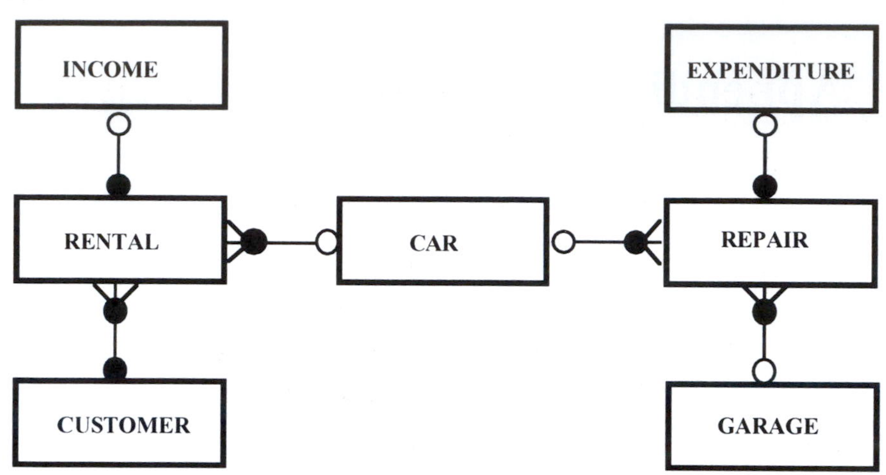
INCOME
EXPENDITURE
RENTAL
CAR
REPAIR
CUSTOMER
GARAGE

2.5

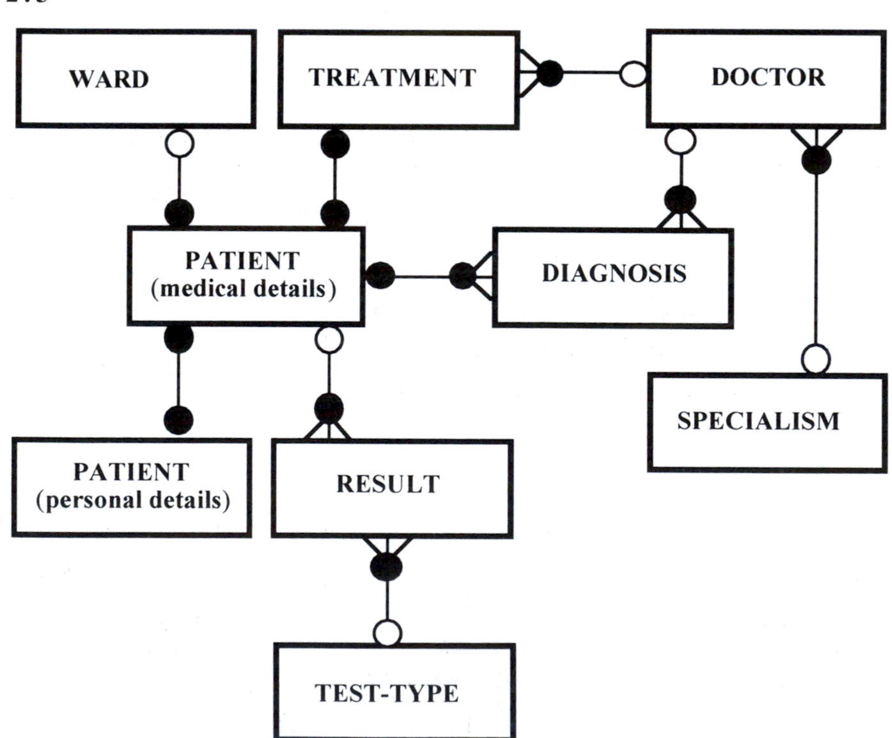
WARD
TREATMENT
DOCTOR
PATIENT
(medical details)
DIAGNOSIS
PATIENT
(personal details)
RESULT
SPECIALISM
TEST-TYPE

2.6

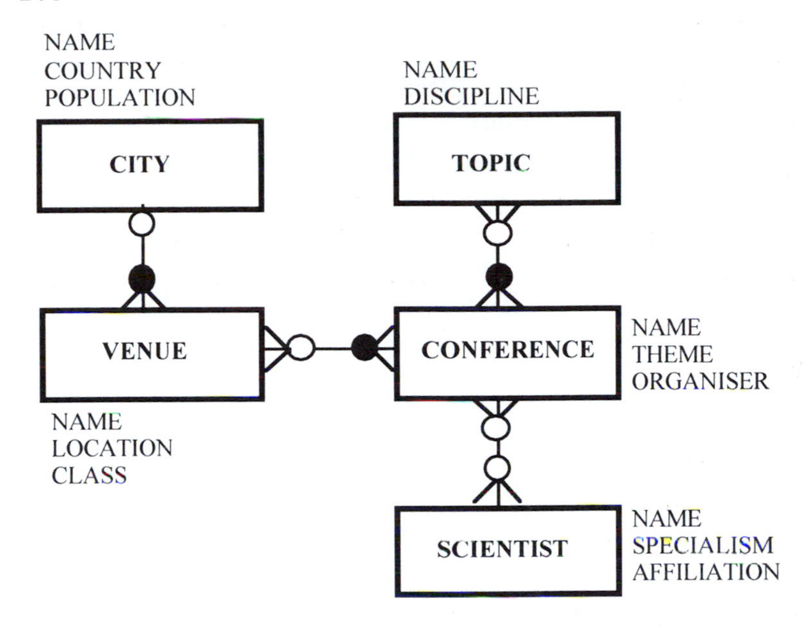
NAME
COUNTRY
POPULATION
CITY
NAME
DISCIPLINE
TOPIC
VENUE
CONFERENCE
NAME
THEME
ORGANISER
NAME
LOCATION
CLASS
SCIENTIST
NAME
SPECIALISM
AFFILIATION

2.7

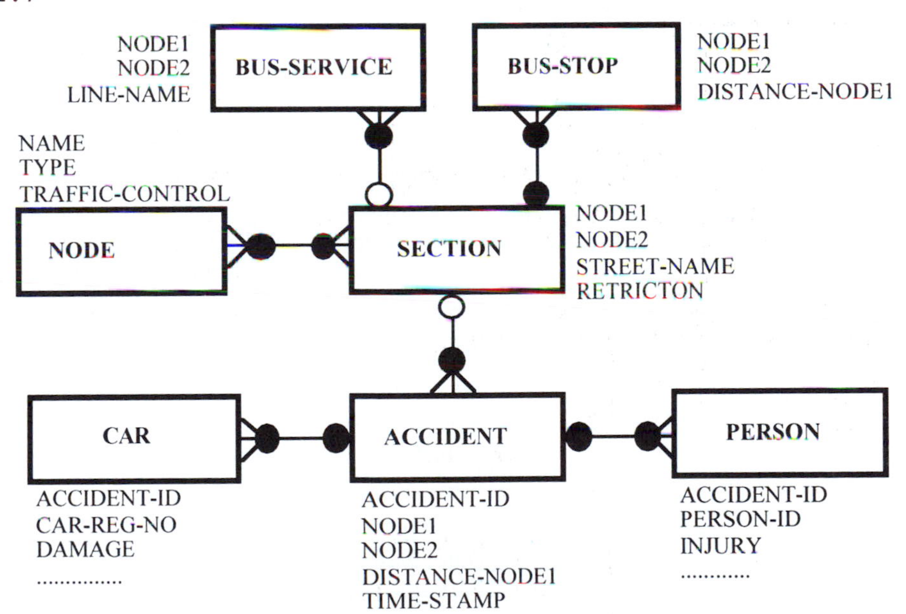
NODE1
NODE2
LINE-NAME
BUS-SERVICE
BUS-STOP
NODE1
NODE2
DISTANCE-NODE1
NAME
TYPE
TRAFFIC-CONTROL
NODE
SECTION
NODE1
NODE2
STREET-NAME
RETRICTON
CAR
ACCIDENT
PERSON
ACCIDENT-ID
CAR-REG-NO
DAMAGE
....
ACCIDENT-ID
NODE1
NODE2
DISTANCE-NODE1
TIME-STAMP
ACCIDENT-ID
PERSON-ID
INJURY
....

CHAPTER 3

For 2.2

MEMBER (M-Id, Name, Address, Since, Recommender)
ENGINEER (E-Id, Name, Grade, Car-Id)
REPAIR (Date, Reg-No, M-Id, E-Id, Fault, Repair, Cost, Surcharge)

For 2.3

DEPARTMENT (Dept-Code, Dept-Name, Manager-Id)
PROJECT (Project-Id, Title, Start-Date, Due-Date, Cost, PManager-Id)
EMPLOYEE (Employee-Id, Name, Job-Title, Dept-Code)
SKILLS (Employee-Id, Skill-Name, Proficiency)
TEAM (Employee-Id, Employee-Role, Project-Id)

For 2.4

INCOME (Amount, Pay-Date, Source-Type, Source-Id, Car-Reg#)
EXPEND (Amount, Pay-Date, Destination-Type, Destination-Id, Car-Reg#)
CAR (Reg#, Make, Model, Prod-Year, Engine-Size, Fuel-Type,
Passenger#, Price, Purchase-Date, Rent-Price, Insurance#, ...)
RENTAL (Rent#, Reg#, Date-Out, Date-In, Customer-Driver#, Deposit)
REPAIR (Rep#, Reg#, Date-Out, Date-Back, Garage-Id, *other details*)
CUSTOMER (Customer-Driver#, Status, Name, Address, Telephone, CredCard#)
GARAGE (Garage-Id, Garage-Name, Address, Pay-Arrangements, *other details*)

If INCOME.Source-Type = '**Rental**' then INCOME.Source-Id is defined on the same domain as the RENTAL.Rent#, otherwise INCOME.Source-Id is set to be equal to the relevant document identifier. Similarly, EXPEND.Destination-Id is equal to either some REPAIR.Rep# or to the relevant payment request identifier.

For 2.5

WARD (Ward-Name, Capacity, Building-Name, ...)
PATIENT (Patient-Id, GP-Id, Doctor-Id, Admission-Reason, Admission-Date)
PERSONAL-DETAILS (Patient-Id, Name, Address, Birth-Date, NHS#,...)
TREATMENT (Patient-Id, *treatment details*, Doctor-Id, Start-Date)
DOCTOR (Doctor-Id, Name, Specialism-Code, Position, ...)
SPECIALISM (Specialism-Code, *descriptive details*)
DIAGNOSIS (Patient-Id, Doctor-Id, Date, *diagnosis details*)
RESULTS (Patient-Id, Test-Code, Date-Taken, *details*)
TEST-TYPE (Test-Code, *descriptive details*)

CHAPTER 4

4.1

(a) **project select** BOOK (PName = 'Pitman')(Title)

(b) **project join (project select** BOOK (Pname = 'MIT Press')) (AName),
AUTHOR: [AName = AName]) (Specialism)

(c) **project join (project select** BOOK (Title = 'A guide to DB2')) (PName),
PUBLISHER: [PName = PName]) (Location)

(d) **project divide**
(**project** BOOK (AName, PName), D : [AName | AName]) (PName)
where D(AName) is a relation containing two tuple-values *Smith* and *Jones*.

(e) **project join (select join (project** BOOK (AName, PName), PUBLISHER :
[PName = PName]) (Location = 'Paris'),
AUTHOR: [AName = AName]) (AName, Address)

4.2

(a) **project join (project select** DEPARTMENT (DeptName = 'R/D') (DeptNo),
EMPLOYEE : [DeptNo = DeptNo]) (EmpNo, EmpName)

(b) **project join (project select (join**
(**project select** DEPARTMENT (Location = 'Geneva') (DeptNo, ManagerNo),
EMPLOYEE : [DeptNo = DeptNo]) (EmpName = 'Smith')
(EmpNo,ManagerNo), EMPLOYEE : [EmpNo = ManagerNo]) (EmpName)

(c) **project join (project select** DEPARTMENT (Location = 'New York')
(DeptNo), EMPLOYEE : [DeptNo = DeptNo) (EmpNo, EmpName, Salary)

4.3 Denoting T = TAPE, M = MEMBER, B = BORROWING

(a) **project** T(Title)

(b) X2 := **project** (**select** T(Title= 'Quadrophenia') (Catalogue#)

(c) X2 \ (**project** (**select join** (X2, B: [Catalogue# = Catalogue#])
(Return-Date = 'Not-Yet') (Catalogue#)

(d) R1 := **project join** (**project select** T(Title = 'Paradise Lost') (Catalogue#),

B:[Catalogue# = Catalogue#])) (Id)

R := **project join**(R1, M: [Id = Id]) (Id, Name)

(e) **project** T(Title) \ (**project select join**(T, B: [Catalogue# =Catalogue#])
(Borrow-Date + 730 > *sysdate*) (Title))

4.4

The proof schema is as follows:

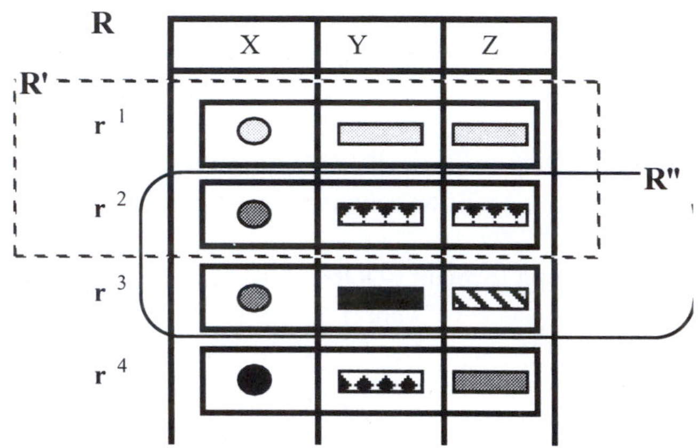

R' = **restrict** R(Y = Z), R'' = **select** R ($X = x_0$). Since R' ∩ R'' = R'' ∩ R'
restrict (**select** R($X = x_0$)) (Y = Z)) ≡ **select** (**restrict** R(Y = Z)) ($X = x_0$))

4.5

(a)

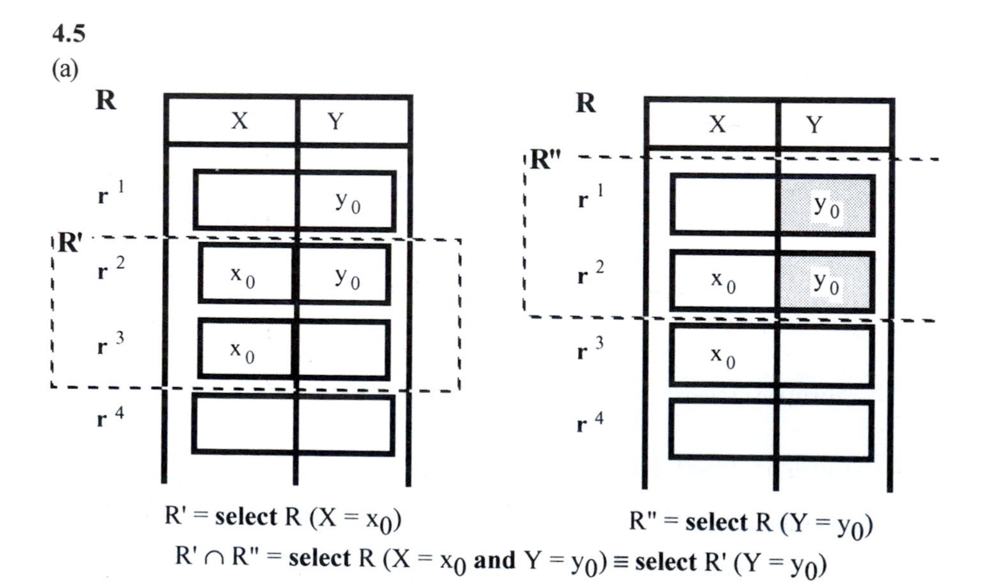

R' = **select** R ($X = x_0$) R'' = **select** R ($Y = y_0$)

R' ∩ R'' = **select** R ($X = x_0$ **and** $Y = y_0$) ≡ **select** R' ($Y = y_0$)

(b) **select** R ($X = x_0$ **or** $Y = y_0$) ≡ **select** R ($X = x_0$) ∪ **select** R ($Y = y_0$)

4.6 Assuming the following denotations: dividend R(X,Y), divisor S(Z), P = project R(X): **divide** (R,S : [X | Z]) ≡ P \ (**project** ((P x S) \ R) (X))

4.8

A circular linked list whose records hold the values of **q**-pointers could be used instead thus preventing potential errors arising from the static nature of the array, which does not cater for the cases when the number of records to be joined exceeds the array size. The operations on the indices of the array would have to be replaced by the corresponding actions involving moving cursor (list-pointer).

CHAPTER 6

6.3

(a) Primary key: **STID**

RESIDENCE is in 2NF (all attributes atomic and a single attribute key)

Functional dependencies:

HALL → LOCATION, (HALL, ROOM#) → TYPE

(HALL, RMTYPE) → RATE, (HALL, MEAL) → MC

and all non-prime attributes functionally dependent on STID.

Decomposition results:

(STID, STNAME, FIELD, HOME_AD, HALL, ROOM#, MEAL)

(HALL, LOCATION)

(HALL, MEAL, MC)

(HALL, ROOM#, RMTYPE)

(HALL, RMTYPE, RATE)

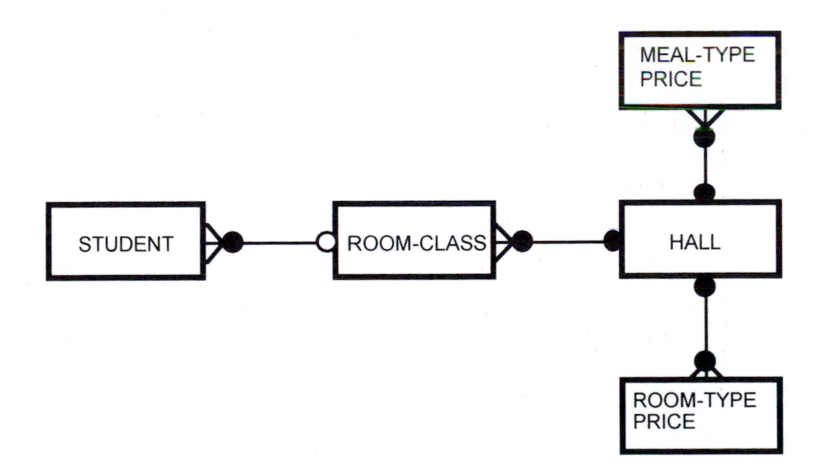

(b) Primary key: **STID**

Functional dependencies:

HALL → LOCATION, TYPE → RATE, MEAL → CHARGE

(HALL, ROOM#) → RMTYPE (and hence - by transitivity - to RATE)

and, as previously, all non-prime attributes functionally dependent on STID

Decomposition results:

(ST#, STNAME, FIELD, HOME_AD, HALL, ROOM#, MEAL)

(HALL, LOCATION)

(MEAL, CHARGE)

(HALL, ROOM#, RMTYPE)

(RMTYPE, RATE)

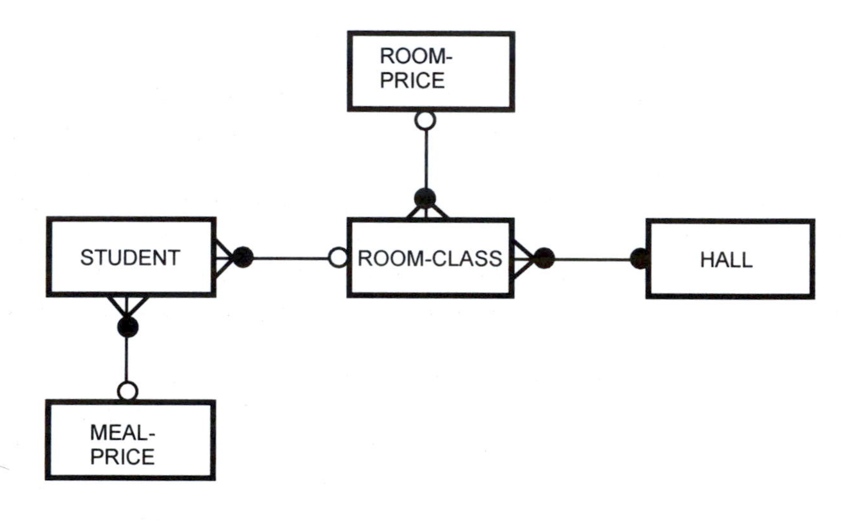

6.4

BCNF definition requires that all determinants in a relation are (candidate) keys; thus eliminates functional dependencies that are only defined on part of the keys, which is the definition of the second normal form.

CHAPTER 7

7.2

(a) The original proof from Fagin (1977) is given below (with minor modifications).

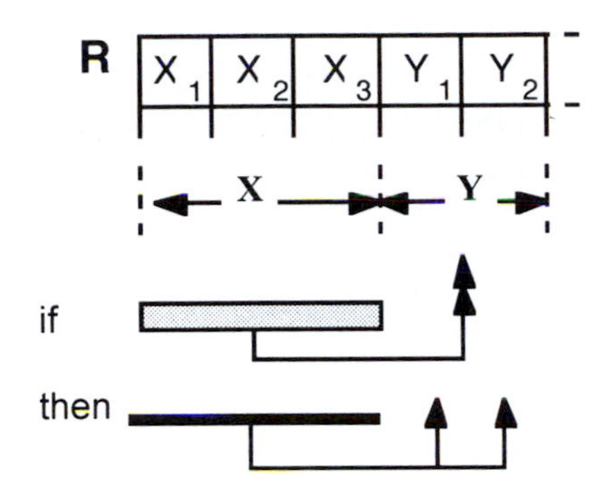

for **R** to be in 4NF

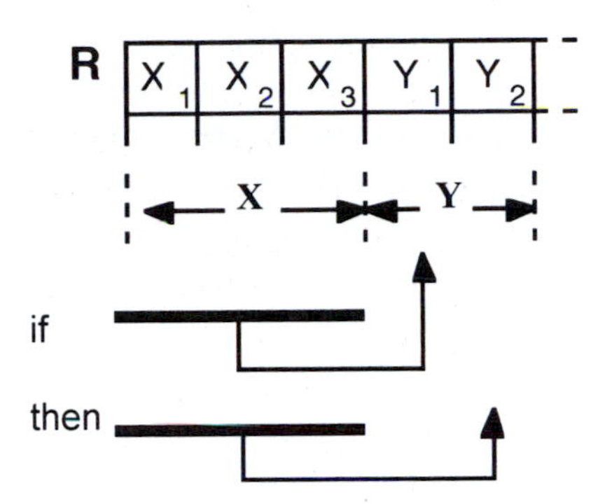

for **R** to be in BCNF

Assume that **R** is in 4NF but not in BCNF. Since **R** is not in BCNF, there is a functional dependency $X \rightarrow Y$ which holds for **R**, and there is an attribute A such that $X \rightarrow A$ does not hold for **R**. Let $Y' = Y \setminus X$. Since $X \rightarrow Y$ holds for **R**, so does $X \rightarrow Y'$. Since X and Y' are disjoint, Y' is nonempty and X and Y' do not partition the attributes of **R**; then the nontrivial multivalued dependency $X \rightarrow\rightarrow Y'$ also holds for **R**. By the definition of 4NF, since $X \rightarrow\rightarrow Y'$ holds for **R**, so does $X \rightarrow A$. We derived a contradiction; hence the original assumption is false.

(b) Assume that R (KEY, A, B) is in BCNF but not in 3NF. Then R would have to violate the definition of 3NF in either not being in 2NF (which is impossible by 6.4) or having at least one attribute transitively dependent on the key. Transitivity would require either $A \rightarrow B$ or $B \rightarrow A$ thus making at least one of the attributes a determinant - which is prevented by the definition of BCNF.

CHAPTER 8

For 4.1

(a) **select** Title **from** BOOK **where** PName = 'Pitman'

(b) **select** Specialism **from** BOOK, AUTHOR
where BOOK.AName = AUTHOR.AName
and BOOK.PName = 'MIT Press'

(c) **select** Location **from** BOOK, PUBLISHER
where BOOK.PName = PUBLISHER.PName
and BOOK.Title = 'A guide to DB2'

(d) (**select** PName **from** BOOK **where** AName = 'Smith') **intersect**
(**select** PName **from** BOOK **where** AName = 'Jones')

(e) **select** AName, Address **from** BOOK, PUBLISHER, AUTHOR
where BOOK.PName = PUBLISHER.PName
and PUBLISHER. A.Name = AUTHOR. AName
and Location = 'Paris' **and** Title = 'A guide to DB2'

For 4.2

(a) **select** EmpNo, EmpName **from** DEPARTMENT, EMPLOYEE
where DEPARTMENT. DeptNo = EMPLOYEE. DeptNo
and DeptName = 'R/D'

(b) **select** EmpName **from** EMPLOYEE **where** EmpNo **in**
(**select** ManagerNo **from** DEPARTMENT, EMPLOYEE
where DEPARTMENT. DeptNo = EMPLOYEE. DeptNo
and Location = 'Geneva' **and** EmpName = 'Smith')

(c) **select** EmpNo, EmpName, Salary **from** DEPARTMENT, EMPLOYEE
where EMPLOYEE.DeptNo = DEPARTMENT.DeptNo
and Location = 'New York'

For 4.3

(a) **select distinct** Title **from** T

(b) **select** Catalogue# **from** T **where** Title = 'Quadrophenia'

(c) (**select** Catalogue# **from** T **where** Title = 'Quadrophenia') **minus**
(**select** T.Catalogue# **from** T, B **where** T.Catalogue# = B.Catalogue#
and Title = 'Quadrophenia' **and** Return-Date = 'Not-Yet')

(d) **select distinct** M.Id, Name **from** T, B, M **where** Title = 'Paradise Lost'
and T.Catalogue# = B.Catalogue# **and** B.Id = M.Id

(e) **select distinct** Title **from** T) **minus**
(**select distinct** Title **from** T, B
where T.Catalogue# = B.Catalogue# **and** Borrow-Date + 730 > *sysdate*)

Appendix B: Denotations, logic, sets

DENOTATIONS

$\in$	belongs to
$\cup$	union
$\cap$	intersection
x	Cartesian product
\	difference
$\subset$	inclusion
$\subseteq$	inclusion or equality
$\equiv$, **iff**	if and only if, identity
$\forall$	**for all** (universal quantifier)
$\exists$	**there exists** (existential quantifier)
	entity
	attribute
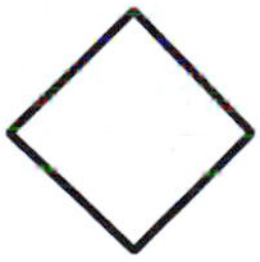	relationship

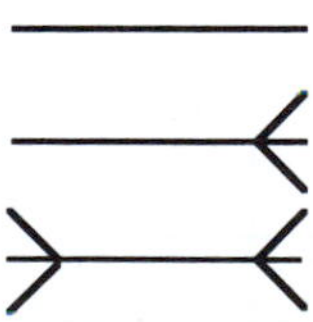

one - to - one

one - to - many

many - to - many

type of relationship

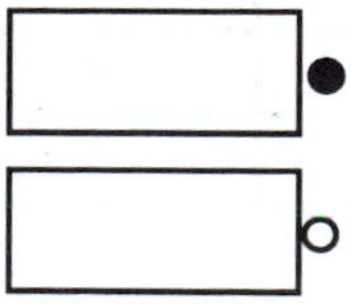

compulsory

optional

participation in a relationship

Relation components

R	A	B	C	D
r^1	r^1_1	r^1_2	r^1_3	r^1_4
r^2	r^2_1	r^2_2	r^2_3	r^2_4
	...	...	...	...
r^n	r^n_1	r^n_2	r^n_3	r^n_4

R,P,Q	relation names (**bold** capital letters)
A, B, C	attribute names (capital Latin letters)
δ, γ	keys (in tuple terms)
Δ, Γ	keys (in attribute name terms)
$\rightarrow$	determines (functional dependency)
$\rightarrow\rightarrow$	multidetermines
JD*	join dependency
φ	dependency set
φ^+	closure of dependency set

ALGEBRA OF PROPOSITIONS

Let **p** and **q** denote statements whose logical values are either true or false but not both. Then, if **p** is true then **not p** is false and vice versa. Compound statements can be formed by combining the simple (sub)statements through the logical connectives, such as **and**, **or**, **not**, **if ... then ...**, **iff.**
The logical values of the statements so composed are defined as follows:

p	q	NEGATION **not** p	CONJUNCTION p **and** q	DISJUNCTION p **or** q	IMPLICATION **if** p **then** q	EQUIVALENCE p **iff** q
F	F	T	F	F	T	T
F	T	T	F	T	T	F
T	F	F	F	T	F	F
T	T	F	T	T	T	T

Some propositions are always true irrespective of the logical values of their components. Such propositions are called tautologies. Similarly, a proposition whose logical value is always false is called a contradiction. Here is a summary of the most common laws of the algebra of propositions:

Double negation	**not not** p ≡ p
Idempotence	p **or** p ≡ p, p **and** p ≡ p
Associativity	(p **or** q) **or** r ≡ p **or** (q **or** r) (p **and** q) **and** r ≡ p **and** (q **and** r)
Commutativity	p **or** q ≡ q **or** p, p **and** q ≡ q **and** p
Distributivity	p **or** (q **and** r) ≡ (p **or** q) **and** (p **or** r) p **and** (q **or** r) ≡ (p **and** q) **or** (p **and** r)
Identity	p **or** false ≡ p, p **and** true ≡ p p **or** true ≡ true, p **and** false ≡ false
Complement	p **or not** p ≡ true, p **and not** p ≡ false **not** true ≡ false, **not** false ≡ true
De Morgan's Laws	**not** (p **or** q) ≡ **not** p **and not** q **not** (p **and** q) ≡ **not** p **or not** q

SET OPERATIONS

A set is a well-defined collection of objects, that is every object that belongs to a particular set satisfies certain requirements. Thus, a usual way of defining a set is $A = \{x: P(x) \text{ is true}\}$ where P denotes a proposition on x. An alternative way of defining a set is to list all its elements, e.g. $B = \{0, 1\}$. We write $x \in A$ to say that an element x belongs to A.

Sets may wholly include one another. A set X is a subset of a set Y, which is denoted by $X \subset Y$, if every element of X is also an element of Y. All sets under investigation (i.e. in a particular application) are assumed to be subsets of the Universal Set, typically denoted by U. On the other hand, an empty set {}, that is a set which contains no elements, is a subset of every other set.

POWER SET

For a given set A, a class of all subsets of A (including the empty set and the set A itself) is called a power set of A and usually denoted by P(A). If the number of elements of A is n, then the number of elements of P(A) is 2^n.

OPERATIONS ON SETS

Union	$X \cup Y = \{a: a \in X \textbf{ or } a \in Y\}$
Intersection	$X \cap Y = \{a: a \in X \textbf{ and } a \in Y\}$
Difference	$X \setminus Y = \{a: a \in X \textbf{ and not } a \in Y\}$
Complement	$X^c = U \setminus X = \{a: a \in U \textbf{ and not } a \in X\}$

PRODUCT SETS

The Cartesian product of two sets A and B consists of all ordered pairs $<a, b>$ where $a \in A$ and $b \in B$

$$A \times B = \{<a, b>: a \in A \textbf{ and } b \in B\}$$

RELATION

A relation R from a set A to a set B is a subset of $A \times B$:

$$R \subset A \times B \equiv R(a,b) \equiv a\, R\, b \equiv\, <a, b> \in R$$

FUNCTION

A relation R is a function if the following conditions are satisfied:

$$\forall a \in A, \exists b \in B: R(a,b)$$

$$\forall a \in A, \forall b_1, b_2 \in B: R(a, b_1) \textbf{ and } R(a, b_2) \textbf{ iff } R(a) = b$$

Bibliography

J.R. Abrial, Data Semantics, *Database Management*, Klimbie & Koffeman (eds), IFIP Working Conference on Database Management, North Holland, 1974

W.W. Armstrong, *Dependency Structures of Database Relationships*, Proceedings of 1974 IFIP Congress, North Holland, Amsterdam, 1974

C.Beeri, P.A. Bernstein, N. Goodman, *A Sophisticate's Introduction to Database Normalization Theory*, Proceedings of 4th Conference on Very Large Data Bases, 1978

P. Beynon-Davies, *Database Systems* (2nd edn), Palgrave, 2000

S.J. Cannan, G.A. Otten, *SQL - The Standard Handbook*, McGraw-Hill, London, 1993

A. Cardenas, *Database Management Systems*, (2nd edn), Allyn & Bacon, Boston, 1985

P.P. Chen, The Entity-Relationship Model: Towards a Unified View of Data, *ACM TODS*, Vol. 1 No 1, September 1976

E.F. Codd, A Relational Model for Large Shared Data Banks, *CACM*, Vol. 13 No 6, June 1970

E.F. Codd, Extending the Database Relational Model to Capture More Meaning, *ACM TODS,* Vol 4. No 4, December 1979

E.F. Codd, The 1981 ACM Turing Award Lecture: Relational Database - A Practical Foundation for Productivity, *CACM*, Vol. 25 No 2, February 1982

T. Connolly *et al.*, *Database Systems - A Practical Approach to Design Implementation and Management*, Addison Wesley, 1996

D.D. Chamberlin, R.F. Boyce, *SEQUEL: A Structured English Query Language*, Proceedings of ACM SIGMOD Workshop on Data Description, Ann Arbor, Michigan, 1974

D.D. Chamberlin *et al.*, A History and Evaluation of System R, *CACM*, Vol. 24 No 10, October 1981

C.J .Date, *Relational Database - Selected Readings*, Addison Wesley, 1986

C.J. Date, *A Guide to the SQL Standard*, Addison Wesley, 1987

C.J. Date, H Darwen, *Foundation for Object/Relational Databases - The Third Manifesto*, Addison Wesley, 1998

C.J. Date, *An Introduction to Database Systems* (7th edn), Addison Wesley, Reading, Massachusetts, 2001

B.P. Desai, *An Introduction to Database Systems*, West Publishing Co, New York, 1990

R. Elmasri, S. Navathe, *Fundamentals of Database Systems*, (3rd edn), Benjamin/Cummings Publishing Co, Reading, Massachusetts, 2000

R. Fagin, Multivalued Dependencies and a New Normal Form for Relational Databases, *ACM TODS*, Vol. 2 No 3, September 1977

R. Fagin, *Normal Forms and Relational Database Operators*, Proceedings of the ACM SIGMOD International Conference on Management of Data, 1979

J.J. van Griethuysen, *Concepts and Terminology for the Conceptual Schema and the Information Base*, ISO TC97/SC5/WG3-N695, 1982

W. Harris, *Databases for Business Users*, Pitman, London, 1992

I.J. Heath, *Unacceptable File Operations in a Relational Database*, Proceedings of the 1971 ACM SIGFIDET Workshop on Data Description, Access and Control, November 1971

W. Kent, *Data and Reality*, North Holland, 1978

W. Kent, Consequences of Assuming a Universal Relation, *ACM TODS*, Vol. 6 No 4, December 1981

H.F. Korth, A. Silberschatz, *Database System Concepts*, McGraw-Hill, New York, 1991

D.M. Kroenke, *Database Processing*, (4th edn), Macmillan, New York, 1992

R.F. van der Lans, *The SQL Standard - A Complete Reference*, Prentice Hall, 1989

B. Laurie, P. Laurie, *Apache: The Definitive Guide*, O'Reilly, 1999

J. Melton, A.R. Simon, *Understanding the New SQL*, Morgan Kaufmann, 1993

D. Maier, *The Theory of Relational Databases*, Computer Science Press, Rockville, Maryland, 1983

D. Maier, J.D Ullman, M.Y. Vardi, On the Foundations of the Universal Relation Model, *ACM TODS*, Vol. 9 No 2, June 1984

F.R. McFadden, J.A. Hoffer, *Database Management* (3rd edn), Addison Wesley, 1991

R. McFadyen, V. Kanabar, *An Introduction to Structured Query Language*, WCB, 1991

P. O'Neil, E. O. Neil, *Database - Principles, Programming and Performance*, (2nd edn), Morgan Kaufmann Publishers, 2001

*SQL*Plus‰ User's Guide & Reference*, Oracle Corporation, 1997

PL/*SQL‰ User's Guide & Reference*, Oracle Corporation, 1992

C. Ricardo, *Database Systems, Principles, Design & Implementation*, Macmillan, 1990

J. Rissanen, *Theory of Joins for Relational Databases - A Tutorial Survey*, Proceedings of 7th Symposium of Mathematical Foundations of Computer Science, Springer-Verlag, Berlin, Germany, 1979

N. Ryan, D. Smith, *Database Systems Engineering*, International Thomson Computer Press, 1995

R. Schwartz, T. Christiansen, L. Wall, *Learning Perl*, O'Reilly, 1997

E. Sciore, A Complete Axiomatization for Full Join Dependencies, *JACM*, Vol. 29 No 2, April 1982

S. Stanczyk, *Programming in SQL*, Pitman, London 1991

M. Stonebraker (ed.), *Readings in Database Systems*, Morgan Kaufmann, 1988

D.C. Tsichritzis, A. Klug, The ANSI/X3/SPARC DBMS Framework: Report of the Study Group on Database Management Systems, *Information Systems*, Vol. 3, 1978

D.C. Tsichritzis, F.H. Lochovsky, *Data Models*, Prentice-Hall, Englewood Cliffs, New Jersey, 1982

J.D. Ullman, *Principles of Database Systems*, (2nd edn), Computer Science Press, 1982

J.D. Ullman, J Widom, *A First Course in Database Systems*, Prentice Hall, 1997

L. Wall, T. Christiansen, J. Orwant, *Programming Perl*, O'Reilly, 2000

R. Yarger, G. Reese, T. King, *MySQL and mSQL*, O'Reilly, 1999

C. Zaniolo, *Analysis and Design of Relational Schemata for Database Systems*, PhD Thesis,Technical Report UCLA-ENG-7669, University of California, Los Angeles, July 1976

Index